Quintics with Symmetries

Solving Some Solvable Classes of Polynomials of Degree 5

Achim Plum

Quintics with Symmetries

Solving Some Solvable Classes of Polynomials of Degree 5

Formulas and Derivations

Impressum

Bibliografische Information der Deutschen Nationalbibliothek: Die Deutsche Nationalbibliothek verzeichnet diese Publikation in der Deutschen Nationalbibliografie; detaillierte bibliografische Daten sind im Internet über dnb.dnb.de abrufbar.

Quintics with Symmetries
Third Edition

Autor: Achim H. Plum
Herstellung und Verlag: BoD – Books on Demand, Norderstedt

ISBN: 9-783755-714798

Content

Foreword

It has been known for a long time that there is no general analytical solution to determine zeros of quintics (polynomials of degree 5) by radical expressions. However, certain classes of quintic equations can be solved, including both reducible and irreducible cases.

In this book, a wide range of reducible quintics is analysed which can be linked to symmetric settings of zeroes. Specifically, if there are two square roots, or if more than two zeroes of a quintic equation are located in a circular symmetric fashion - be it centrally around 0 or an arbitrary point in the field of complex numbers - this corresponds, with the exception of a 5th roots scenario, to a reducible quintic. Solutions are provided by derived equations of lower degree than 5 which are presented in this book. In addition, various examples for quintic equations that fulfil the respective conditions are shown in detail. Knowledge about the solvability of polynomial equations of degree 4 or lower is being leveraged.

It should be noted that I did not find publicly available resources providing a comparable compilation of formulas. Errors in the text or mathematical content cannot be completely excluded; any feedback is highly appreciated. Also, the content of this book is limited with respect to the mathematical theory like for example Galois Theory; in the meantime, I have also published an e-book elaborating more on that with title *Galois Groups and Field Extensions for Solvable Quintics*.

In the reference section, existing resources are provided which have been studied to ensure an appropriate problem statement and context. Many of those resources are Wikipedia documents which contain some elaborated content on certain aspects, and often provide many additional links to other Wikipedia pages with related content or relevant reference materials.

By the book, I would like to share the collected results in the problem domain. The motivation has been to find a significant subset of (reducible) quintics which can be described by certain symmetric conditions of their zeroes leading to radical expressions. Consequently, quintics which do not fall into one of the solvable categories as presented here must have a non-symmetric positioning of the zeroes regarding a cyclic symmetry.

In this third edition, the foreword has been rewritten to present the content of this book in the context of reducible quintics as its scope. In addition, printing errors as well as calculation errors in a few examples – found and verified with the *SageMath* software system – have been corrected.

I would like to thank my wife Gisela and my children for their support.

September, 2022
Achim Plum
Author

1. Introduction

A general solution for resolving polynomials of degree 5 called quintics by radicals (root expressions) is not possible as shown by Niels Henrik Abel in 1824 (after initial work by Paolo Ruffini in 1799), and algebraically understood in Galois theory, for which the foundation was laid by Évariste Galois shortly after in the 19th century. [1]

Nevertheless, certain classes of quintics can be solved by resolvents; these are among those which have a Galois group which is solvable. The question is though how such quintics can be identified. From earlier work [2] it is clear that

a) The zeroes of a quintic leave their traces within the coefficients of the polynomials;
b) Certain symmetry patterns and conditions may exist in quintics.

All of the following considerations are based on the field $\mathbb{C}$ of complex numbers, which is algebraically closed. Among others, the following categories of quintics are solvable by roots:

- Those which contain roots as zeroes;
- Those which contain multiple (3, 4 or 5 times) zeroes;
- Those which contain a root symmetry for zeroes not centred around $0 \in \mathbb{C}$.

Furthermore, a resolvent with two variables is provided for all quintics. As expected, the occurrence and dependency of two variables corresponds to the fact that a quintic cannot be generally solved. However, with additional information or conditions, cases might be identified which actually are solvable by means of the resolvent.

Beyond this, a heuristic approach is provided starting with an analysis of the squares of zeroes and of the constant value to potentially find solutions.

In this work, all resolvents are derived from the unchanged coefficients of the quintic; no Tschirnhaus [5] or any other transformation is applied to it.

1.1 Solvable Scenarios

In principle, all scenarios which show a circular symmetry of some of the zeroes are solvable. First of all, it is shown that all quintics which contain roots as zeroes can be resolved. A *root* in this book is called a number x which fulfils the equation

$$x^n = a$$

for an integer $n > 1$ called degree and some number $a \neq 0$ of $\mathbb{C}$. Since the subject of this work are quintics, roots can have degrees 2,3,4, or 5. The above equation is equivalent to

$$x^n - a = 0$$

with its left term being a factor of the polynomial.

The second group of resolvable quintics are those which have multiple zeroes with multiplicity 3,4 or 5. In these cases, a binomial pattern is left within the coefficients of the polynomial, which can be used to derive resolvents. An arbitrary zero $\neq 0$ of multiplicity 2 would be a case of the asymmetric scenario which cannot be generally solved.

The third resolvable class are quintics with \`offset roots´. This expression is meant to describe a symmetry between at least three zeroes which does not have its centre at $0 \in \mathbb{C}$ though. By an offset transformation, solutions can be derived from corresponding quintics with roots.

It should be noted that a quintic can belong to several classes whereof some are more specific than others. As a simple example, a quintic with a 4-times zero of course also does have a 3-times zero.

1.2 Prerequisites and Conventions

Prerequisites for reading are a certain level of understanding of the field of complex numbers $\mathbb{C}$, the exponential and trigonometric functions, and polynomials as elements of $\mathbb{C}[X]$, the algebraic ring of polynomials over the field $\mathbb{C}$ which allows the Euclidian division (the division of polynomials with remainder). Furthermore, some familiarity with resolving quadratic, cubic and quartic equations is helpful. References for further reading are provided within the text and at the end.

Throughout this book, the following conventions are used:

- Polynomials are written in the form
  $$f(x) = \sum_{i=0}^{5} k_i x^i;\ g_j(x) = \sum_{i=0}^{j} l_i x^i;\ k_i, l_i \in \mathbb{C}; j < 5$$
  i.e. the polynomial which is subject of the analysis is called f, and other polynomials with a lower degree than 5 are usually called g_j.
- It is always assumed in quintics f that $k_5 = 1$, which means that the polynomial is normalized. Likewise, in g_j, $l_j = 1$ is assumed if not specified differently. The k_i, l_i are elements of $\mathbb{C}$ or of $\mathbb{R}$ if specifically mentioned.
- It is always assumed in f that $k_0 \neq 0$ because otherwise 0 would be known as a zero.
- The variables $a, b, c\ d, e \neq 0$ are used as constant numbers of $\mathbb{C}$. These - or their negatives - are also often used to identify zeroes. $n, n_i \neq 0$ are also used to denote zeroes.
  Constants used in polynomial linear factors like in
  $f(x) = (x - a)(x - b)(x - c)(x - d)(x - e)$
  or as zeroes of f are $\neq 0$ by the nature of the topic, since otherwise 0 would be a zero of f, which is a trivial case.
- In diagrams, the complex field $\mathbb{C}$ is shown as a circle and the axis of real numbers. The purpose of diagrams is to illustrate concepts only; they are figurative and are sketched based on an arbitrary selection of important variables.
- Since in general roots (radicals) of complex numbers in $\mathbb{C}$ are not unique, special attention might be necessary in scenarios which contain complex (non-real) roots. Without further analysis, it is assumed that normally the principal value of a root serves the purpose; if that turns out not to be accurate in some cases, deeper analyses of roots may be required.

2 Quintics with Roots

This chapter describes how quintics with roots can be resolved. For this class of quintics, the target is to determine one or several zeros which are roots.

A fashionable way to express solutions are exponential function terms of imaginary numbers, which are located on a circle around the centre 0 in $\mathbb{C}$ in case they have the same absolute value (for some context, see also [10] for an article on cyclotomic polynomials). This should be a sufficient representation for roots, which closely correlates to the characteristics of the exponential / trigonometric functions which may have transcendental or algebraic values and in particular values which can be expressed by radical expressions. Simple examples:

$$cos\frac{\pi}{6}=\frac{\sqrt{3}}{2}; sin\frac{\pi}{6}=0.5$$

$$cos\frac{2\pi}{5}=\frac{\sqrt{5}-1}{4}; sin\frac{2\pi}{5}=\frac{\sqrt{10+2\sqrt{5}}}{4}$$

Cf. [3] for a resource which gives some insight into algebraic values and radical expressions of trigonometric functions. Vice versa, it is obvious that if $z \in \mathbb{C}$ can be expressed by radicals, then the absolute value of z can be expressed by radicals as well – applying Pythagoras´ theorem with the real and imaginary parts of
$z := re^{i\alpha} = r \cdot (\cos\alpha + i\sin\alpha)$ for $r \in \mathbb{R}^+, \alpha \in \mathbb{R}$. [9] Example with $\alpha = \frac{\pi}{3}$:

$$z=\sqrt{3}e^{\frac{\pi}{3}i}=\sqrt{3}(\frac{1}{2}+\frac{\sqrt{3}}{2}i)=\frac{\sqrt{3}}{2}+\frac{3}{2}i \Rightarrow |z|=\sqrt{\frac{3}{4}+\frac{9}{4}}=\sqrt{3}$$

2.1 Existence of a 5th Root

How to recognize:	$f(x) = x^5 + k_0$ $k_4 = k_3 = k_2 = k_1 = 0$

With $a := -k_0$ the polynomial can be written as

$f(x) := x^5 - a$

Case analysis:

<table>
<tr>
<td>

For $a \in \mathbb{R}, a > 0$:

$r := \sqrt[5]{a}$

Zeroes: $n_k = re^{\frac{2\pi ik}{5}}; k = 0, \dots, 4$

$n_4 = \overline{n_1};\ n_3 = \overline{n_2};\ n_0 = r$

Diagram:

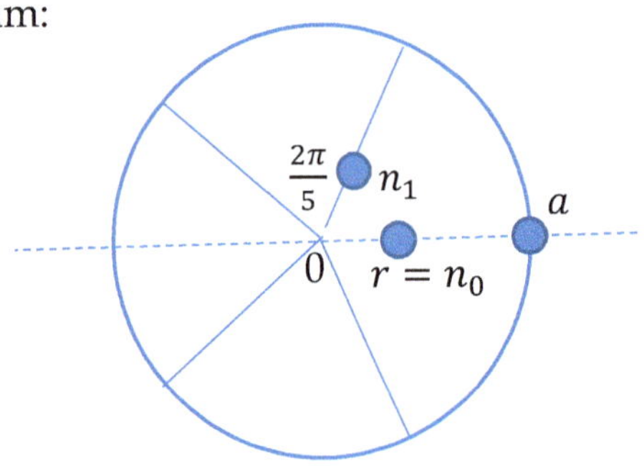

</td>
<td>

For $a \in \mathbb{R}, a < 0$:

$r := \sqrt[5]{-a}$

Zeroes: $n_k = re^{\frac{2\pi i(2k+1)}{10}}; k = 0, \dots, 4$

$n_4 = \overline{n_0};\ n_3 = \overline{n_1};\ n_2 = -r$

Diagram:

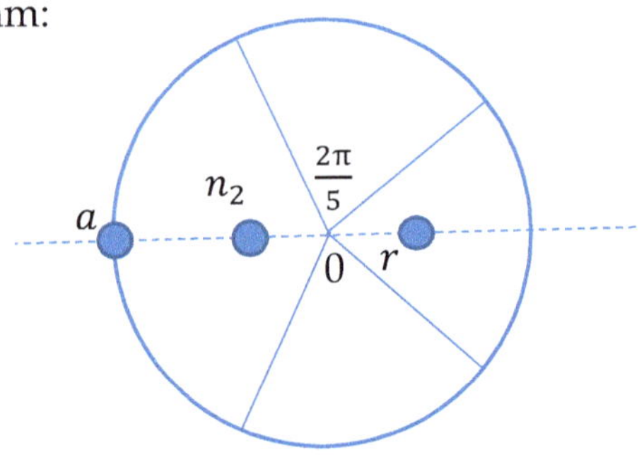

</td>
</tr>
<tr>
<td>

For $k_0 \notin \mathbb{R}$: $a \notin \mathbb{R}$:

$r := \sqrt[5]{|a|}$; $\exists \alpha \in \mathbb{R}$ with

$n_0 = re^{i\alpha}, n_0^5 = a = r^5 e^{i \cdot 5\alpha}$

Zeroes: $n_k = re^{i\left(\alpha + \frac{2\pi k}{5}\right)}; k = 0, \dots, 4$

Diagram:

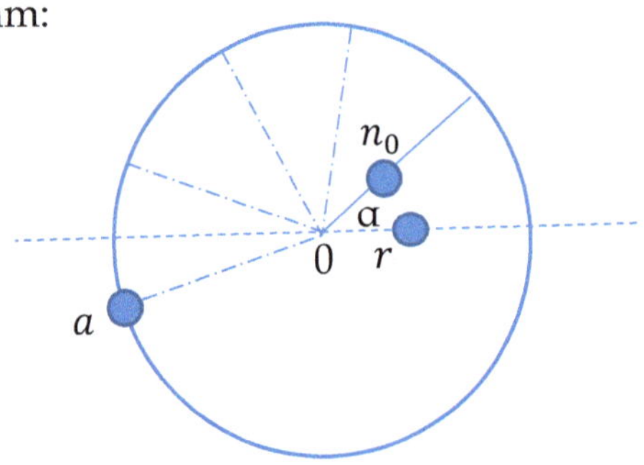

</td>
<td>

Examples:

For $a \in \mathbb{R}, a > 0$:

$f(x) = x^5 - 2$

$r = \sqrt[5]{2};\ a = 2; n_0 = \sqrt[5]{2}; n_1 = \sqrt[5]{2} \cdot e^{\frac{2\pi i}{5}}$

$= \sqrt[5]{2} \cdot \left(\frac{\sqrt{5}-1}{4} + i \cdot \frac{\sqrt{2}}{4}\sqrt{5+\sqrt{5}}\right)$

For $a \in \mathbb{R}, a < 0$:

$f(x) = x^5 + 2$

$r = \sqrt[5]{2};\ a = -2; n_2 = -\sqrt[5]{2}$

For $a \notin \mathbb{R}$:

$f(x) = x^5 - 2e^{i\frac{10}{9}\pi}$

$r = \sqrt[5]{2}; a = 2e^{i\frac{10}{9}\pi}; \alpha = \frac{2}{9}\pi;\ n_0 = \sqrt[5]{2}e^{i\frac{2}{9}\pi};$

$n_1 = \sqrt[5]{2}e^{i(\frac{2}{9}+\frac{2}{5})\pi} = \sqrt[5]{2}e^{i\frac{28}{45}\pi}$

</td>
</tr>
</table>

2.2 Existence of a 4th Root

How to recognize:	$f(x) = x^5 + k_4x^4 + k_1x + k_0$ $k_3 = k_2 = 0$ $k_4 \neq 0 \wedge k_1 \neq 0$ $k_0 = k_1 \cdot k_4$

With $a \coloneqq -k_1$ and $b \coloneqq -k_4$ the polynomial can be written as
$f(x) \coloneqq (x^4 - a)(x - b)$

Case analysis:

For $a \in \mathbb{R}, a > 0$: $r \coloneqq \sqrt[4]{a}$ Zeroes: $n_k = re^{\frac{2\pi ik}{4}}; k = 0, \dots, 3$ $n_0 = r;\ n_2 = -r;\ n_1 = ir;\ n_3 = -ir$ Diagram: 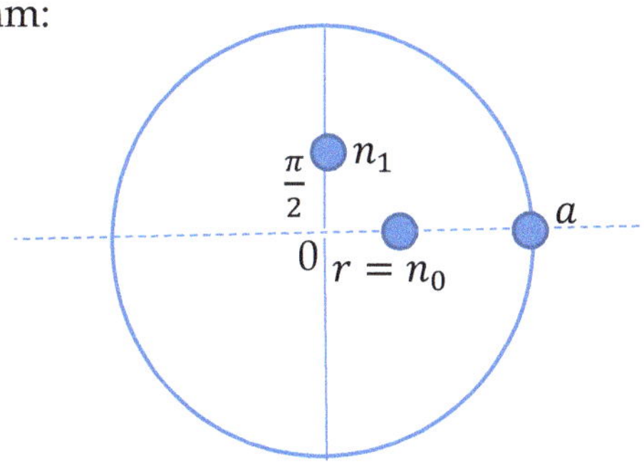	For $a \in \mathbb{R}, a < 0$: $r \coloneqq \sqrt[4]{-a}$ Zeroes: $n_k = re^{\frac{2\pi i(2k+1)}{8}}; k = 0, \dots, 3$ $n_3 = \overline{n_0};\ n_2 = \overline{n_1}$ Diagram: 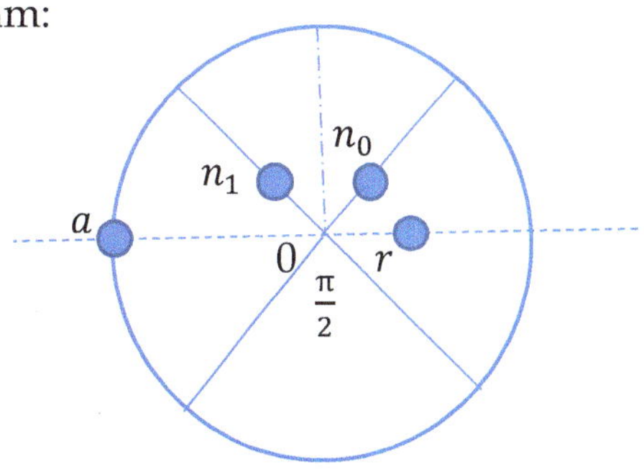
For $k_0 \notin \mathbb{R}, a \notin \mathbb{R}$: $r \coloneqq \sqrt[4]{\lvert a \rvert};\ \exists \alpha \in \mathbb{R}$ with $n_0 = re^{i\alpha}, n_0^4 = a = r^4e^{i \cdot 4\alpha}$ Zeroes: $n_k = re^{i\left(\alpha + \frac{2\pi k}{4}\right)}; k = 0, \dots, 3$ Diagram: 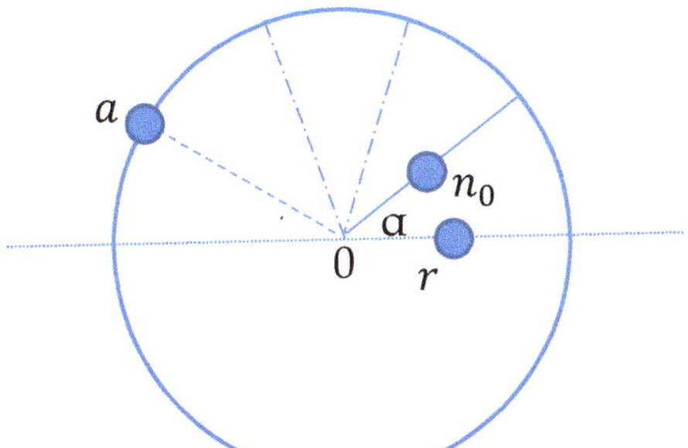	Examples: For $a \in \mathbb{R}, a > 0$: $f(x) = x^5 - 3x^4 - 2x + 6 = (x^4 - 2)(x - 3)$ $r = \sqrt[4]{2};\ a = 2; n_0 = \sqrt[4]{2}$ For $a \in \mathbb{R}, a < 0$: $f(x) = x^5 - 3x^4 + 2x - 6 = (x^4 + 2)(x - 3)$ $r = \sqrt[4]{2};\ a = -2; n_0 = 2e^{i\frac{1}{4}\pi}$ For $a \notin \mathbb{R}$: $f(x) = (x^4 - 2e^{i\frac{8}{9}\pi})(x - 3)$ $r = \sqrt[4]{2}; a = 2e^{i\frac{8}{9}\pi}; \alpha = \frac{1}{4}\pi;\ n_0 = \sqrt[4]{2}e^{i\frac{2}{9}\pi};$ $n_1 = \sqrt[4]{2}e^{i(\frac{2}{9}+\frac{1}{2})\pi} = \sqrt[4]{2}e^{i\frac{13}{18}\pi}$

2.3 Existence of a 3rd Root and a Square Root

How to recognize:	$f(x) = x^5 + k_3x^3 + k_2x^2 + k_0$ $k_4 = k_1 = 0$ $k_3 \neq 0 \wedge k_2 \neq 0$ $k_0 = k_2 \cdot k_3$

With $a := -k_2$ and $b := -k_3$ the polynomial can be written as
$f(x) := (x^3 - a)(x^2 - b)$

Zeroes are the roots of the two factors. Refer to the next section for a comprehensive view on the 3rd root.

Example: $f(x) = x^5 + 5x^3 - 2x^2 - 10$ $f(x) = (x^3 - 2)(x^2 + 5)$

2.4 Existence of a 3rd Root Without a Square Root

How to recognize:	$f(x) = x^5 + k_4x^4 + k_3x^3 + k_2x^2 + k_1x + k_0$ $k_4 \neq 0 \wedge k_3 \neq 0 \wedge k_2 \neq 0$ $k_1 = k_2 \cdot k_4$ $k_0 = k_2 \cdot k_3$ $k_4 \neq 0$ because otherwise case in *section* 2.3.

With $a := -k_2$ and $b, c := -\frac{1}{2}k_4 \pm \sqrt{\frac{k_4^2}{4} - k_3}$ the polynomial can be written as
$f(x) := (x^3 - a)(x - b)(x - c)$

Case analysis:

<table>
<tr>
<td>
For $a \in \mathbb{R}, a > 0$:
$r := \sqrt[3]{a}$
Zeroes: $n_k = re^{\frac{2\pi ik}{3}}; k = 0, \dots, 2$
$n_0 = r;\ n_2 = \overline{n_1}$
Diagram:

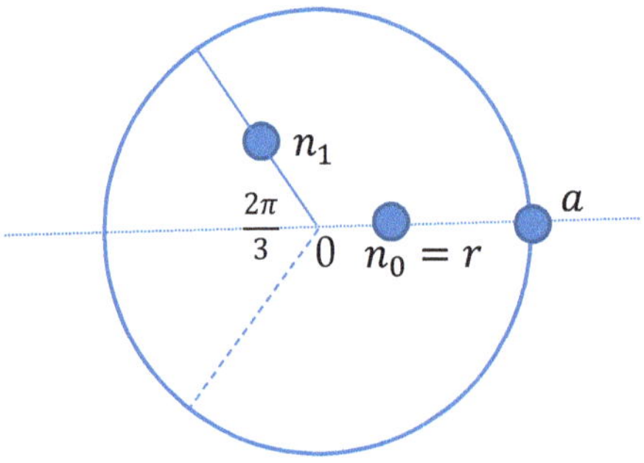

</td>
<td>
For $a \in \mathbb{R}, a < 0$:
$r := \sqrt[3]{-a}$
Zeroes: $n_k = re^{\frac{2\pi i(2k+1)}{6}}; k = 0, \dots, 2$
$n_2 = \overline{n_0};\ n_1 = -r$
Diagram:

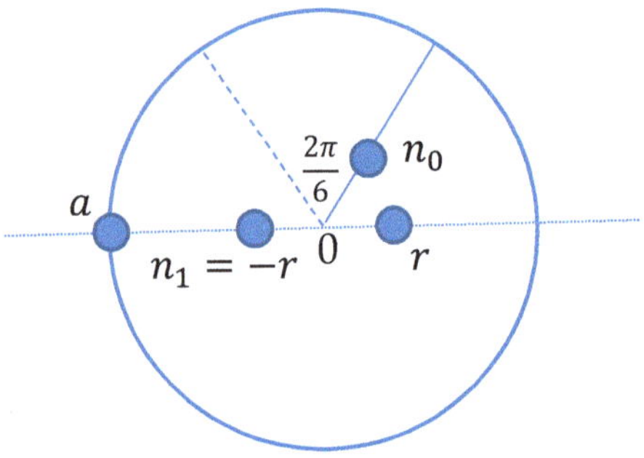

</td>
</tr>
</table>

For $k_0 \notin \mathbb{R}$, $a \notin \mathbb{R}$:

$r := \sqrt[3]{|a|}$; $\exists \alpha \in \mathbb{R}$ with

$n_0 = re^{i\alpha}$, $n_0^3 = a = r^3 e^{i \cdot 3\alpha}$

Zeroes: $n_k = re^{i\left(\alpha+\frac{2\pi k}{3}\right)}$; $k = 0, \dots, 2$

Diagram:

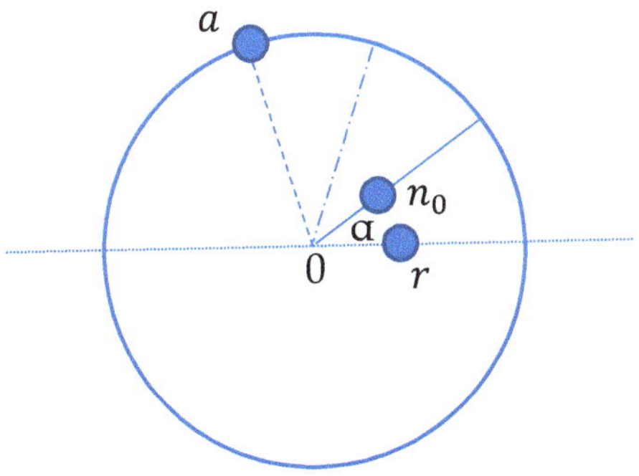

Examples:

For $a \in \mathbb{R}, a > 0$:

$f(x) = x^5 + \frac{9}{2}x^4 - \frac{5}{2}x^3 - 2x^2 - 9x + 5$

$f(x) = (x^3 - 2)(x + 5)(x - \frac{1}{2})$

For $a \in \mathbb{R}, a < 0$:

$f(x) = x^5 + \frac{9}{2}x^4 - \frac{5}{2}x^3 + 2x^2 + 9x - 5$

$f(x) = (x^3 + 2)(x + 5)(x - \frac{1}{2})$

For $a \notin \mathbb{R}$:

$f(x) = x^5 + \frac{9}{2}x^4 - \frac{5}{2}x^3 - 2e^{i\frac{2}{3}\pi}x^2 - 9e^{i\frac{2}{3}\pi}x$

$+5e^{i\frac{2}{3}\pi}$

$f(x) = (x^3 - 2e^{i\frac{2}{3}\pi})(x + 5)(x - \frac{1}{2})$

Derivation:

$f(x) = x^5 - (b + c)x^4 + bcx^3 - ax^2 + a(b + c)x - abc$

$b + c = -k_4 \wedge b + c = -\frac{k_1}{k_2} \Rightarrow k_4 = \frac{k_1}{k_2}$

$bc = k_3 \wedge bc = \frac{k_0}{k_2} \Rightarrow k_3 = \frac{k_0}{k_2}$

$k_1 = ab + ac = -k_2 b - k_2 \frac{k_3}{b}$

$\Rightarrow k_1 b = -k_2 b^2 - k_2 k_3$

$b = -\frac{1}{2}\frac{k_1}{k_2} \pm \sqrt{\frac{1}{4}\frac{k_1^2}{k_2^2} - k_3} = -\frac{1}{2}k_4 \pm \sqrt{\frac{k_4^2}{4} - k_3}$

2.5 Existence of Two Square Roots

This scenario is also covered by the next section 2.6 with special condition $d = -b$. However, if two square roots do exist, the expressions are much simpler due to the additional symmetry. How to recognize:	$f(x) = x^5 + k_4x^4 + k_3x^3 + k_2x^2 + k_1x + k_0$ $\exists a, b, c \in \mathbb{C}$ $k_4 = -c$ $k_3 = -(a+b)$ $k_2 = c(a+b) = k_4 \cdot k_3$ $k_1 = ab$ $k_0 = -abc = k_1 \cdot k_4$

With $a := -\frac{k_3}{2} \pm \sqrt{\frac{k_3^2}{4} - k_1}$, $b := -k_3 - a$ and $c := -k_4$ the polynomial can be written as $f(x) := (x^2 - a)(x^2 - b)(x - c)$ for one of the two possible a values.

Case analysis:

For $\sqrt{a}, \sqrt{b} \in \mathbb{R}$:

Examples:

$f(x) = x^5 - 2x^4 - \frac{7}{3}x^3 + \frac{14}{3}x^2 + \frac{2}{3}x - \frac{4}{3}$

$\Rightarrow f(x) = (x^2 - \frac{1}{3})(x^2 - 2)(x - 2)$

$f(x) = x^5 - 3x^4 - \frac{7}{3}x^3 + 7x^2 + \frac{2}{3}x - 2$

$\Rightarrow f(x) = (x^2 - \frac{1}{3})(x^2 - 2)(x - 3)$

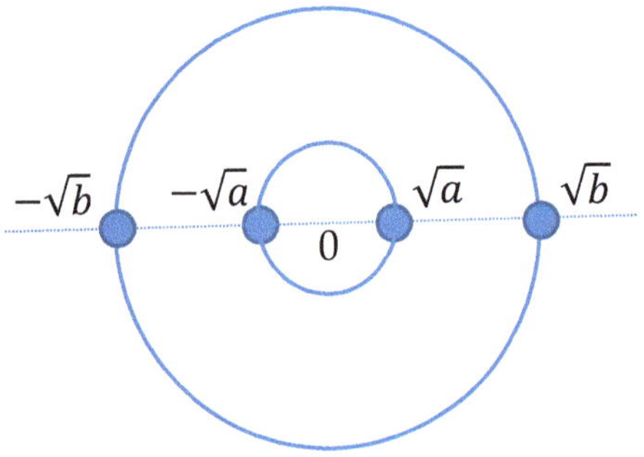

For $\sqrt{a} \notin \mathbb{R}, \sqrt{b} \in \mathbb{R}$:

Example:

$f(x) = x^5 - 2x^4 - \frac{5}{3}x^3 + \frac{10}{3}x^2 - \frac{2}{3}x + \frac{4}{3}$

$\Rightarrow f(x) = (x^2 + \frac{1}{3})(x^2 - 2)(x - 2)$

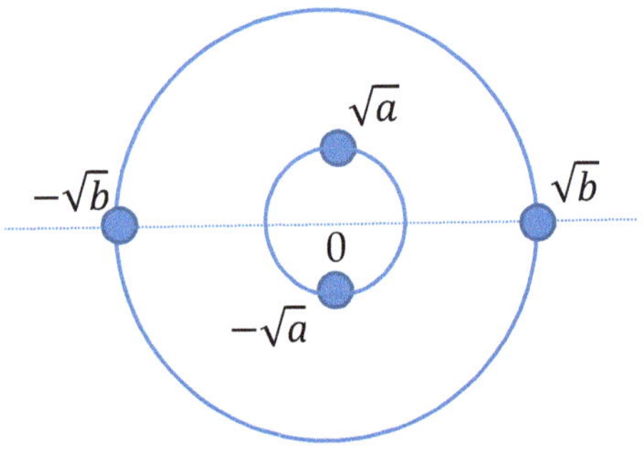

$\sqrt{a}, \sqrt{b} \notin \mathbb{R}$ (Diagram: arbitrary case):

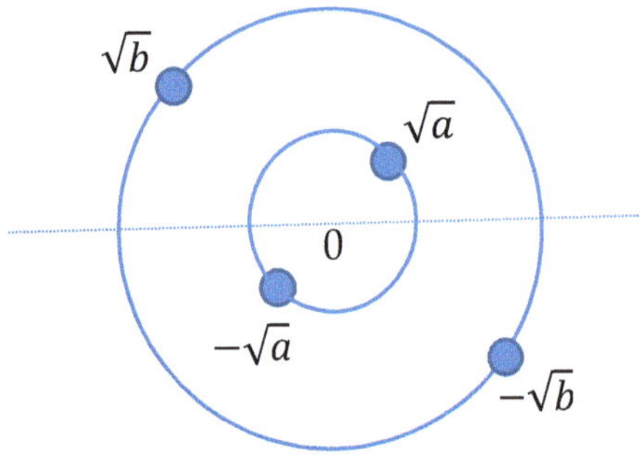

Example:

$f(x) = x^5 - 2x^4 + \frac{7}{3}x^3 - \frac{14}{3}x^2 + \frac{2}{3}x - \frac{4}{3}$

$\Rightarrow f(x) = (x^2 + \frac{1}{3})(x^2 + 2)(x - 2)$

Derivation:

The scenario can be considered symmetric in a and b. Let be $a, b, c \neq 0$, otherwise 0 would be a zero.

$f(x) := (x^2 - a)(x^2 - b)(x - c)$

$= (x^4 - (a+b)x^2 + ab)(x - c)$

$= x^5 - cx^4 - (a+b)x^3 + c(a+b)x^2 + abx - abc$

$\Rightarrow k_4 = -c; k_3 = -(a+b)$

$k_2 = c(a+b) = k_4k_3; k_1 = ab; k_0 = -abc$

$\Rightarrow c = -k_4; b = \frac{k_1}{a}$

$a = -k_3 - \frac{k_1}{a} \Rightarrow a^2 + k_3a + k_1 = 0$

With $a = -\frac{k_3}{2} \pm \sqrt{\frac{k_3^2}{4} - k_1}$

solutions for c, a, b are found.

2.6 Existence of One Square Root

How to recognize:	$f(x) = x^5 + k_4x^4 + k_3x^3 + k_2x^2 + k_1x + k_0$ $\exists a, b, c, d \in \mathbb{C}, a \neq 0$ $k_4 = -(b + c + d)$ $k_3 = (bc + bd + cd - a)$ $k_2 = a(b + c + d) - bcd = -ak_4 - \frac{bcd}{a}$ $k_1 = -a(bc + bd + cd)$ $k_0 = abcd$ If $k_4 = 0$, the equations are simplified to: $k_2 = -bcd$ $k_0 = -ak_2$ If in addition $k_2 = 0$, then also $k_0 = 0$, and the polynomial has 0 as a zero. Therefore, it can be assumed that $k_2 \neq 0$.

If $k_4 = 0$, then $a = -\frac{k_0}{k_2}$ is the square root of the quintic, and

if $k_4 \neq 0$, then one of $a = -\frac{k_2}{2k_4} \pm \sqrt{\frac{k_2^2}{4k_4^2} - \frac{k_0}{k_4}}$ is the square root, such that the quintic can be written as

$f(x) := (x^2 - a)(x - b)(x - c)(x - d)$ for one of the two possible a values.

The other zeroes b, c and d can be determined by determining the zeroes of the polynomial of degree 3 received by polynomial division $f(x)\colon (x^2 - a)$.

Case analysis:

For $k_4 = 0$:	For $k_4 \neq 0$:
Example: $f(x) = x^5 - \frac{15}{2}x^3 - 6x^2 + \frac{7}{2}x + 3$ $a = \frac{3}{6} = \frac{1}{2}$ $(x^5 - \frac{15}{2}x^3 - 6x^2 + \frac{7}{2}x + 3)\colon (x^2 - \frac{1}{2})$ $= x^3 - 7x - 6$ $= (x + 1)(x + 2)(x - 3)$	Example: $f(x) = x^5 - x^4 - \frac{21}{2}x^3 - \frac{15}{2}x^2 + 5x + 4$ $a = -\frac{15}{4} \pm \sqrt{\frac{225}{16} + 4} = -\frac{15}{4} \pm \frac{17}{4}$ $= \frac{1}{2}$ *or* -8; Division for $a = \frac{1}{2}$: $(x^5 - x^4 - \frac{21}{2}x^3 - \frac{15}{2}x^2 + 5x + 4)\colon (x^2 - \frac{1}{2})$ $= x^3 - x^2 - 10x - 8$ $= (x + 1)(x + 2)(x - 4)$

Derivation:

$f(x) := (x^2 - a)(x - b)(x - c)(x - d)$
$= (x^2 - a)(x^3 - (b + c + d)x^2 + (bc + bd + cd)x - bcd)$
$= x^5 - (b + c + d)x^4 + (bc + bd + cd)x^3 - bcdx^2$
$-ax^3 + a(b + c + d)x^2 - a(bc + bd + cd)x + abcd$
$\Rightarrow k_4 = -(b + c + d); k_3 = bc + bd + cd - a;$
$k_2 = -ak_4 - bcd; k_1 = -a(bc + bd + cd);$
$k_0 = abcd$
$\Rightarrow -ak_4 = k_2 + bcd$

Case 1: $k_4 \neq 0$: $a = -\frac{k_2+bcd}{k_4} \wedge bcd = \frac{k_0}{a} \Rightarrow a = -\frac{k_2}{k_4} - \frac{k_0}{k_4 a}$ $\Rightarrow a^2 + \frac{k_2}{k_4}a + \frac{k_0}{k_4} = 0$ $\Rightarrow a = -\frac{k_2}{2k_4} \pm \sqrt{\frac{k_2^2}{4k_4^2} - \frac{k_0}{k_4}}$	Case 2: $k_4 = 0 \wedge k_2 \neq 0$: $b + c + d = 0 \Rightarrow k_2 = -bcd \Rightarrow k_0 = -ak_2$ $\Rightarrow a = -\frac{k_0}{k_2}$
Diagram: 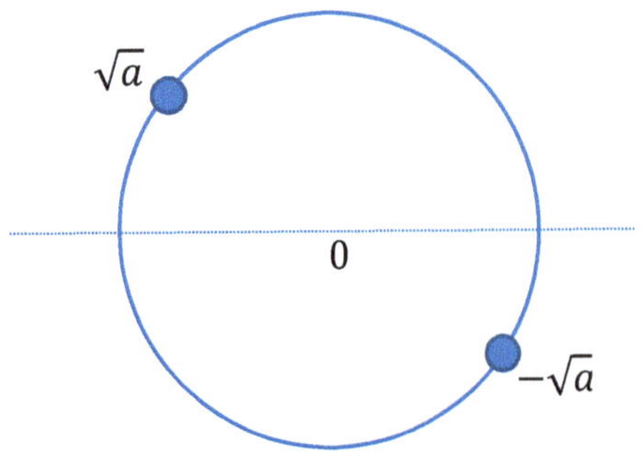	

3 Quintics with Multiple Zeroes

In this chapter quintics are described which have multiple zeroes which occur 5, 4 or 3 times. The main pattern which occurs here is the existence of binomial coefficients as factors of the quintic´s coefficients.

These cases can be seen as special cases of quintics with offset roots (see chapters 4.1 - 4.3) in the sense that the circle of symmetry would have a radius of 0. However, the approach with multiple zeroes is documented for completeness. In the case of 5-times and 4-times zeroes, with $a = 0$ the resolvents are basically the same or comparable to the general ones. In the 3-times zero case, the resolvent for the special case here (chapter 3.3) has a somewhat simpler structure than the one derived in the general case in chapter 4.3.

3.1 Existence of a 5-times Zero

How to recognize:	$f(x) = x^5 + k_4x^4 + k_3x^3 + k_2x^2 + k_1x + k_0$ $\exists a \in \mathbb{C}$ with $a = \frac{k_4}{5}$, such that $f(x) = x^5 + 5ax^4 + 10a^2x^3 + 10a^3x^2$ $+5a^4x + a^5$

By expanding $(x + a)^5$, the binomial coefficients $\binom{5}{j}$ occur as factors in each k_j, potentially with alternating signs in the coefficients of the different powers of x.

In this case, the only zero is $-a$.

Example: $f(x) = x^5 - \frac{10}{3}x^4 + \frac{40}{9}x^3 - \frac{80}{27}x^2 + \frac{80}{81}x - \frac{32}{243}$ $a = -\frac{2}{3}$ $f(x) = \left(x - \frac{2}{3}\right)^5$	

3.2 Existence of a 4-times Zero

How to recognize:	$f(x) = x^5 + k_4x^4 + k_3x^3 + k_2x^2 + k_1x + k_0$ $\exists a, b \in \mathbb{C}$ with $b = k_4 - 4a$, such that $f(x) = (x+a)^4(x+b)$ $= (x^4 + 4ax^3 + 6a^2x^2 + 4a^3x + a^4)(x+b)$ and $k_4 = 4a + b$ $k_3 = 6a^2 + 4ab$ $k_2 = 4a^3 + 6a^2b$ $k_1 = a^4 + 4a^3b$ $k_0 = a^4b$

By expanding $(x+a)^4$ the binomial coefficients $\binom{4}{j-1}$, $j > 0$, and $\binom{4}{j}$ occur as factors of powers of a and b in each k_j, potentially with alternating signs in the coefficients of the different powers of x. They cannot be recognized very easily though due to a shift in powers of a and b.

With $a := \frac{k_4}{5} \pm \sqrt{\frac{k_4^2}{25} - \frac{k_3}{10}}$ and $b := k_4 - 4a$ the quintic can be written as $f(x) := (x+a)^4(x+b)$ for one of the two possible a values.

In this case, the zeroes are $-a, -b$.

Example: $f(x) = x^5 - \frac{2}{3}x^4 - \frac{8}{3}x^3 + \frac{112}{27}x^2 - \frac{176}{81}x + \frac{32}{81}$ $a = -\frac{2}{3}; b := 2$ $f(x) = \left(x - \frac{2}{3}\right)^4 (x+2)$	Example: $f(x) = x^5 + \frac{1}{3}x^4 - \frac{16}{3}x^3 + \frac{184}{27}x^2 - \frac{272}{81}x + \frac{16}{27}$ $a = -\frac{2}{3}; b := 3$ $f(x) = \left(x - \frac{2}{3}\right)^4 (x+3)$

Derivation: $f(x) = (x+a)^4(x+b)$ $= (x^4 + 4ax^3 + 6a^2x^2 + 4a^3x + a^4)(x+b)$ $= x^5 + (4a+b)x^4 + (6a^2 + 4ab)x^3$ $+(4a^3 + 6a^2b)x^2 + (a^4 + 4a^3b)x + a^4b$ $\Rightarrow k_4 = 4a + b \Rightarrow b = k_4 - 4a$ $\Rightarrow k_3 = 6a^2 + 4a(k_4 - 4a) = -10a^2 + 4k_4a$ $\Rightarrow a^2 - \frac{2}{5}k_4a + \frac{k_3}{10} = 0$ $\Rightarrow a = \frac{k_4}{5} \pm \sqrt{\frac{k_4^2}{25} - \frac{k_3}{10}}$

3.3 Existence of a 3-times Zero

How to recognize:	$f(x) = x^5 + k_4x^4 + k_3x^3 + k_2x^2 + k_1x + k_0$ $\exists a, b, c \in \mathbb{C}$, such that $f(x) = (x+a)^3(x+b)(x+c)$ $= (x^3 + 3ax^2 + 3a^2x + a^3)(x+b)(x+c)$ and $k_4 = 3a + b + c$ $k_3 = 3a^2 + 3a(b+c) + bc$ $k_2 = a^3 + 3a^2(b+c) + 3abc$ $k_1 = a^3(b+c) + 3a^2bc$ $k_0 = a^3bc$

In this case, it is more complicated to determine a solution, because there is more combinatorial complexity in the coefficients. However, a zero can be found by a condition which can be expressed by a polynomial of degree 4 or 3, which can be solved by Ferrari´s / Cardano´s equations [6], or by a quadratic equation in simple cases. The found candidates for zeroes are then tested with the quintic and one of them is a zero of it.

We may assume that $a \neq 0$, otherwise 0 would be a triple zero of $f(x)$. In the case that $k_4 = 0$ and $k_3 \neq 0$, a is a solution of the equation

$$a^3 - \frac{3\,k_2}{4\,k_3}a^2 + \frac{5\,k_0}{4\,k_3} = 0$$

or, if $k_3 = 0$, $a^2 - \frac{5\,k_0}{3\,k_2} = 0$ with $k_2 \neq 0$

and for $k_4 \neq 0$:

$$a^4 - \frac{4\,k_3}{3\,k_4}a^3 + \frac{k_2}{k_4}a^2 - \frac{5\,k_0}{3\,k_4} = 0$$

With that, b is a solution of the equation

$$b^2 + (3a - k_4)b + \frac{k_0}{a^3} = 0 \Rightarrow b = \frac{k_4 - 3a}{2} \pm \sqrt{\frac{(k_4-3a)^2}{4} - \frac{k_0}{a^3}}, \text{ and } c = k_4 - 3a - b$$

In these cases, the zeroes are $-a, -b, -c$ for each one of the two possible a and b values.

For the scenario with $k_4 = k_3 = k_2 = 0$, i.e. $f(x) = x^5 + k_1x + k_0$, no solution with a 3-times zero does exist, cf. proof below.

Case analysis:

Example for $k_4 = 0$: $f(x) = x^5 - \frac{17}{3}x^3 + \frac{226}{27}x^2 - \frac{124}{27}x + \frac{8}{9}$ $a = -\frac{2}{3}$; $b = 3$; $c = -1$ $f(x) = \left(x - \frac{2}{3}\right)^3 (x+3)(x-1)$	Example for $k_4 \neq 0$: $f(x) = x^5 - 5x^4 - \frac{8}{3}x^3 + \frac{424}{27}x^2 - \frac{112}{9}x + \frac{80}{27}$ $a = -\frac{2}{3}$; $b = 2$; $c = -5$ $f(x) = \left(x - \frac{2}{3}\right)^3 (x+2)(x-5)$

Derivation:

$f(x) = (x+a)^3(x+b)(x+c) = (x^3 + 3ax^2 + 3a^2x + a^3) \cdot (x^2 + (b+c)x + bc)$
$= x^5 + (3a + b + c)x^4 + (3a^2 + 3a(b+c) + bc)x^3 + (a^3 + 3a^2(b+c) + 3abc)x^2$
$+(a^3(b+c) + 3a^2bc)x + a^3bc$
$\Rightarrow k_4 = 3a + b + c \Rightarrow b + c = k_4 - 3a; k_3 = 3a^2 + 3a(b+c) + bc$
$\Rightarrow a^2 + (b+c)a + \frac{bc-k_3}{3} = 0 \Rightarrow a^2 + (k_4 - 3a)a + \frac{bc-k_3}{3} = -2a^2 + k_4a + \frac{bc-k_3}{3} = 0$
$\Rightarrow a^2 - \frac{k_4}{2}a + \frac{k_3-bc}{6} = 0$

It is $k_0 = a^3bc \Rightarrow bc = \frac{k_0}{a^3}$ with $a \neq 0$, otherwise 0 would be a zero of f

$\Rightarrow a^2 - \frac{k_4}{2}a + \frac{k_3}{6} - \frac{k_0}{6a^3} = 0 \Rightarrow a^5 - \frac{k_4}{2}a^4 + \frac{k_3}{6}a^3 - \frac{k_0}{6} = 0$
Now, also $k_2 = a^3 + 3a^2(b+c) + 3abc \Rightarrow a^3 + 3a^2(k_4 - 3a) + 3\frac{k_0}{a^2} - k_2 = 0$
$\Rightarrow -8a^3 + 3k_4a^2 + 3\frac{k_0}{a^2} - k_2 = 0 \Rightarrow a^5 - \frac{3k_4}{8}a^4 + \frac{k_2}{8}a^2 - \frac{3k_0}{8} = 0$

By subtraction of these two equations of degree 5 in a one gets to
$-\frac{k_4}{8}a^4 + \frac{k_3}{6}a^3 - \frac{k_2}{8}a^2 + \frac{5k_0}{24} = 0$

Scenario $k_4 \neq 0$:
$a^4 - \frac{4k_3}{3k_4}a^3 + \frac{k_2}{k_4}a^2 - \frac{5k_0}{3k_4} = 0$

Scenario $k_4 = k_3 = k_2 = 0$ is not possible for this case:
if $k_4 = 0$, the coefficient terms in a, b, c resulted to $b + c = -3a$
Then, because of $k_3 = k_2 = 0$: eliminating expression bc by $3k_3 - \frac{k_2}{a}$:
$8a^2 + 6a(b+c) = 0 \Rightarrow b + c = -\frac{4}{3}a$
$\Rightarrow -3a = -\frac{4}{3}a$ contradiction because of $a \neq 0$.

Regarding b and c:
$c = k_4 - 3a - b \wedge bc = \frac{k_0}{a^3}$
$\Rightarrow bc = b(k_4 - 3a - b) = \frac{k_0}{a^3} \Rightarrow b^2 - (3a - k_4)b + \frac{k_0}{a^3} = 0$
which can be resolved for b, c.

4 Quintics with Offset Roots

With analytic solutions in place for all quintics with roots as described above, the focus of this chapter is now on quintics with symmetric zeroes but not roots. An example may be that three zeroes are located on a circle in a symmetric way but are not centered around 0 in $\mathbb{C}$. In such a case, one can imagine a quintic which is moved by a linear offset accordingly such that the center of the circle is 0, and then the above techniques for resolvents may be applied to solve the quintic.

Quintics fulfilling this condition are called *Quintics with Offset Roots* in this book. The corresponding quintic with center = 0 is called *the Central Quintic* here, denoted as $f_c(z)$.

It is important to note that, in order to keep complexity manageable, the offset mechanism is only applied to the roots which are subject to the transformation and not to the other zeroes (this is not relevant for chapter 4.1). By this, the polynomial is somewhat deformed when creating the central quintic; however, this is still sufficient to identify the roots.

Also, the fact should be mentioned that for the first case of a 5th offset root (chapter 4.1), a derivation is not included here; the case is rather obvious, since the binomial factors can be clearly isolated.

4.1 Existence of a 5th Offset Root

How to recognize:	$f(x) = x^5 + k_4x^4 + k_3x^3 + k_2x^2 + k_1x + k_0$ $\exists t \in \mathbb{C}$, such that $k_4 = 5t$ $k_3 = 10t^2$ $k_2 = 10t^3$ $k_1 = 5t^4$

Then, $f_c(z) := z^5 + k_0 - t^5 \Rightarrow f(x) = f_c(x + t)$, and for $a := t^5 - k_0$, $\sqrt[5]{a}$ is a solution of $f_c(z) = 0$, with t being the offset distance of $f(x)$ to the central quintic $f_c(z)$. With the substitution $z := x + t$, the central quintic can be transformed back to the offset quintic which has $\sqrt[5]{a} - t$ as a zero.

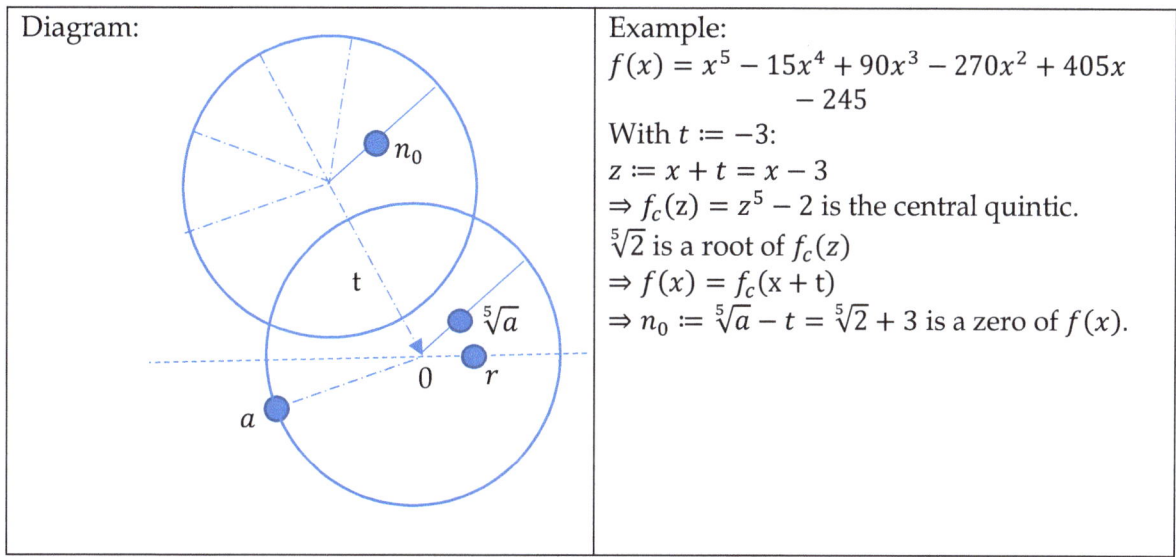

<table>
<tr>
<td>Diagram:
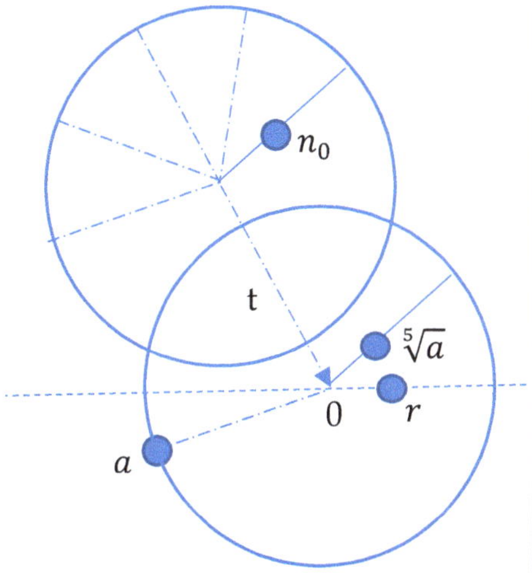
</td>
<td>Example:
$f(x) = x^5 - 15x^4 + 90x^3 - 270x^2 + 405x - 245$
With $t := -3$:
$z := x + t = x - 3$
$\Rightarrow f_c(z) = z^5 - 2$ is the central quintic.
$\sqrt[5]{2}$ is a root of $f_c(z)$
$\Rightarrow f(x) = f_c(x + t)$
$\Rightarrow n_0 := \sqrt[5]{a} - t = \sqrt[5]{2} + 3$ is a zero of $f(x)$.</td>
</tr>
</table>

4.2 Existence of a 4th Offset Root

How to recognize:	$f(x) = x^5 + k_4x^4 + k_3x^3 + k_2x^2 + k_1x + k_0$ $\exists t, a, b \in \mathbb{C}, a, b \neq 0$ such that $k_4 = 4t + b$ $k_3 = 6t^2 + 4tb$ $k_2 = 4t^3 + 6t^2b$ $k_1 = t^4 - a + 4t^3b$

Let $f_c(z) \coloneqq (z^4 - a)(z + b)$. With the given conditions, $f(x) = ((x + t)^4 - a)(x + b)$. Then, $\sqrt[4]{a}$ is a solution of $f_c(z) = 0$, with t being the offset distance (related to the 4th offset root) of $f(x)$ to the central quintic. With the substitution $z \coloneqq x + t$, the factor $(z^4 - a)$ can be transformed back to the corresponding factor of the offset quintic which has $\sqrt[4]{a} - t$ and $-b$ as zeroes.

The solution can be derived by

$$b = k_4 - 4t \Rightarrow k_3 = 6t^2 + 4t(k_4 - 4t) = -10t^2 + 4k_4t$$

$$\Leftrightarrow t^2 - \frac{2k_4}{5}t + \frac{k_3}{10} = 0 \Leftrightarrow t = \frac{k_4}{5} \pm \sqrt{\frac{k_4^2}{25} - \frac{k_3}{10}}$$

One of the two options for t leads to the solution, and with $k_0 = (t^4 - a) \cdot b$, a can be determined accordingly by

$$a = t^4 - \frac{k_0}{b}$$

Diagram: 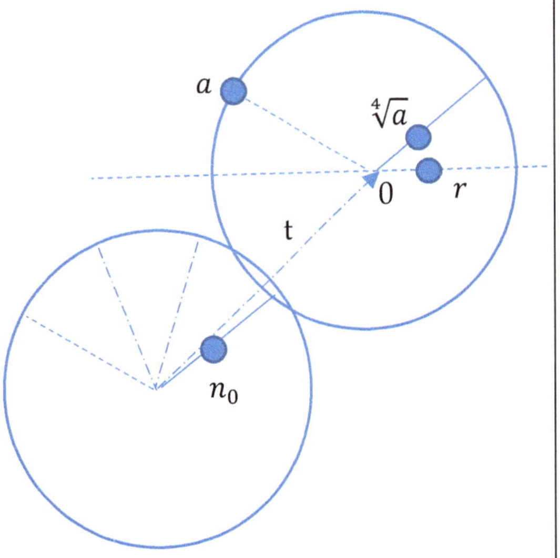	Example: $f(x) = x^5 - 11x^4 + 42x^3 - 54x^2 - 29x + 79$ With $t \coloneqq -3 \Rightarrow b = 1 \Rightarrow a = 2$: $z \coloneqq x + t = x - 3$ $\Rightarrow f_c(z) = (z^4 - 2)(z + 1)$ is the central quintic. $\sqrt[4]{2}$ is a root of $f_c(z)$ $\Rightarrow n_0 \coloneqq \sqrt[4]{a} - t = \sqrt[4]{2} + 3$ is a zero of $f(x) = ((x - 3)^4 - 2)(x + 1)$; $-b = -1$ is a zero of $f(x)$. Example: $f(x) = x^5 - 14x^4 + 78x^3 - 216x^2 + 295x - 158$ With $t \coloneqq -3 \Rightarrow b = -2 \Rightarrow a = 2$: $z \coloneqq x + t = x - 3$ $\Rightarrow f_c(z) = (z^4 - 2)(x - 2)$ is the central quintic. $\sqrt[4]{2}$ is a root of $f_c(z)$ $\Rightarrow n_0 \coloneqq \sqrt[4]{a} - t = \sqrt[4]{2} + 3$ is a zero of $f(x)$; $-b = 2$ is a zero of $f(x)$.

4.3 Existence of a 3rd Offset Root

How to recognize:	$f(x) = x^5 + k_4x^4 + k_3x^3 + k_2x^2 + k_1x + k_0$ $\exists t, a, b, c \in \mathbb{C}$, $a, b, c \neq 0$ such that $k_4 = 3t + (b + c)$ $k_3 = 3t^2 + 3t(b + c) + bc$ $k_2 = t^3 - a + 3t^2(b + c) + 3tbc$ $k_1 = (t^3 - a)(b + c) + 3t^2bc$

Let $f_c(z) \coloneqq (z^3 - a)(z + b)(z + c)$. With the given conditions, $f(x) = ((x + t)^3 - a)(x + b)(x + c)$. Then, $\sqrt[3]{a}$ is a solution of $f_c(z) = 0$, with t being the offset distance (related to the 3rd offset root) of $f(x)$ to the central quintic. With the substitution $z \coloneqq x + t$, the factor $(z^3 - a)$ can be transformed back to the corresponding factor of the offset quintic which has $\sqrt[3]{a} - t$, $-b$ and $-c$ as zeroes.

The values for t, a, b and c are as follows:

t solves the equation $t^4 - \frac{4}{5}k_4t^3 + \frac{2}{15}(k_4^2 + 2k_3)t^2 - \frac{1}{15}(k_2 + k_3k_4)t + \frac{1}{45}(k_2k_4 - k_1) = 0$

$a = 10t^3 - 6k_4t^2 + 3k_3t - k_2$

b solves the equation $(t^3 - a)b^2 - (t^3 - a)(k_4 - 3t)b + k_0 = 0$

$c = k_4 - 3t - b$

Diagram:

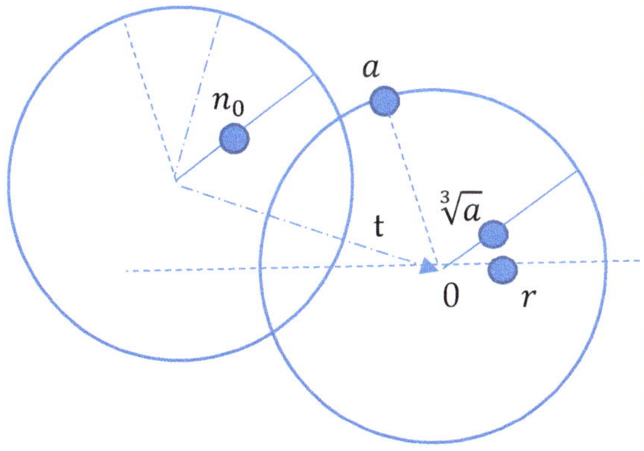

Example:
$f(x) = x^5 - 10x^4 + 34x^3 - 38x^2 - 25x + 58$

With $t \coloneqq -3 \Rightarrow a = 2 \Rightarrow b = 1 \wedge c = -2$:
$z \coloneqq x + t = x - 3$
$\Rightarrow f_c(z) = (z^3 - 2)(z + 1)(z - 2)$ is the central quintic.
$\sqrt[3]{2}$ is a root of $f_c(z)$
$\Rightarrow n_0 \coloneqq \sqrt[3]{a} - t = \sqrt[3]{2} + 3$ is a zero of $f(x) = ((x - 3)^3 - 2)(x + 1)(x - 2)$;
$-b = -1; -c = 2$ are zeroes of $f(x)$.

Derivation:

$f(x) = ((x + t)^3 - a)(x + b)(x + c) = (x^3 + 3tx^2 + 3t^2x + t^3 - a) \cdot (x^2 + (b + c)x + bc)$
$= x^5 + \big(3t + (b + c)\big)x^4 + (3t^2 + 3t(b + c) + bc)x^3 + (t^3 - a + 3t^2(b + c) + 3tbc)x^2$
$\quad + \big((t^3 - a)(b + c) + 3t^2bc\big)x + (t^3 - a)bc$

$\Rightarrow$

$k_4 = 3t + (b + c)$
$k_3 = 3t^2 + 3t(b + c) + bc$
$k_2 = t^3 - a + 3t^2(b + c) + 3tbc$
$k_1 = (t^3 - a)(b + c) + 3t^2bc$
$k_0 = (t^3 - a)bc$

$\Rightarrow$
$b + c = k_4 - 3t$
$bc = k_3 - 3t^2 - 3t(b + c) = k_3 - 3t^2 - 3t(k_4 - 3t) = 6t^2 - 3k_4 t + k_3$
$k_2 = t^3 - a + 3t^2(b + c) + 3tbc = t^3 - a + 3t^2(k_4 - 3t) + 3t(6t^2 - 3k_4 t + k_3)$
$= t^3 - a - 9t^3 + 3k_4 t^2 + 18t^3 - 9k_4 t^2 + 3k_3 t$

$a = 10t^3 - 6k_4 t^2 + 3k_3 t - k_2$

$k_1 = (t^3 - a)(b + c) + 3t^2 bc$
$= \left(t^3 - (10t^3 - 6k_4 t^2 + 3k_3 t - k_2)\right)(k_4 - 3t) + 3t^2(6t^2 - 3k_4 t + k_3)$
$= 27t^4 - 27k_4 t^3 + (6k_4^2 + 9k_3)t^2 - 3(k_2 + k_3 k_4)t + k_2 k_4 + 18t^4 - 9k_4 t^3 + 3k_3 t^2$

$\Leftrightarrow t^4 - \frac{4}{5}k_4 t^3 + \frac{2}{15}(k_4^2 + 2k_3)t^2 - \frac{1}{15}(k_2 + k_3 k_4)t + \frac{1}{45}(k_2 k_4 - k_1) = 0$

The equation in t can be resolved, and with a and t identified, b and c can be determined:
$c = k_4 - 3t - b$
$\Rightarrow k_0 = (t^3 - a)b(k_4 - 3t - b) \Leftrightarrow (t^3 - a)b^2 - (t^3 - a)(k_4 - 3t)b + k_0 = 0$
which can be resolved to determine b.

4.4 Existence of Two Offset Square Roots

How to recognize:	$f(x) = x^5 + k_4x^4 + k_3x^3 + k_2x^2 + k_1x + k_0$ $\exists t, a, b, c \in \mathbb{C}, t, a, b, c \neq 0$ such that $k_4 = 4t + c$ $k_3 = 6t^2 + 4ct - (a + b)$ $k_2 = 4t^3 + 6ct^2 - 2(a + b)t - (a + b)c$ $k_1 = t^4 + 4ct^3 - (a + b)t^2 - 2(a + b)ct$ $+ab$

Let $f_c(z) := (z^2 - a)(z^2 - b)(z + c)$. With the given conditions, $f(x) = ((x + t)^2 - a)((x + t)^2 - b)(x + c)$. Then, $\sqrt{a}$ and $\sqrt{b}$ are solutions of $f_c(z) = 0$, with t being the offset distance (related to the two offset square roots with one unique offset) of $f(x)$ to the central quintic. With the substitution $z := x + t$, the factors $(z^2 - a)$ and $(z^2 - b)$ can be transformed back to the corresponding factors of the offset quintic which has $\sqrt{a} - t$, $\sqrt{b} - t$ and $-c$ as zeroes.

The values for t, a, b and c are as follows:

t solves the equation $t^3 - \frac{3}{5}k_4t^2 + \left(\frac{1}{10}k_4^2 + \frac{1}{20}k_3\right)t + \frac{1}{40}(k_2 - k_4k_3) = 0$

With $c = k_4 - 4t$, a solves the equation
$a^2 + (10t^2 - 4k_4t + k_3)a - 11t^4 - 4(6c - k_4)t^3 - (k_3 - 8k_4c)t^2 - 2k_3ct + k_1 = 0$

$b = -10t^2 + 4k_4t - k_3 - a$

Diagram: 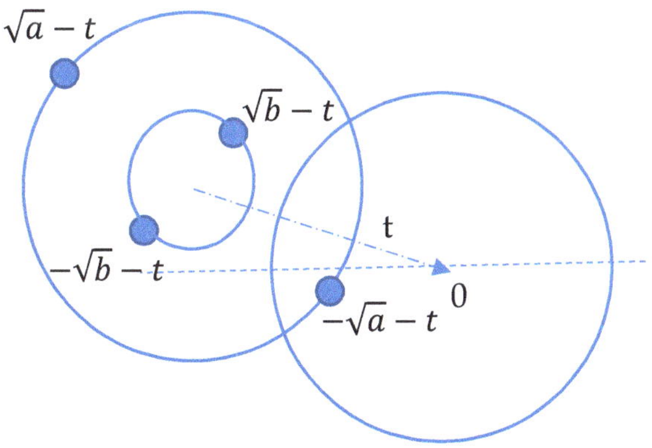 	Example: $f(x) = x^5 - 14x^4 + 77x^3 - 208x^2 + 274x$ $- 140$ With $t := -3 \Rightarrow a = 2 \Rightarrow b = -1 \wedge c = -2$: $z := x + t = x - 3$ $\Rightarrow f_c(z) = (z^2 - 2)(z^2 + 1)(z - 2)$ is the central quintic. $\sqrt{2}$ and i are roots of $f_c(z)$ $\Rightarrow \sqrt{a} - t = \sqrt{2} + 3$ and $\sqrt{b} - t = i + 3$ are zeroes of $f(x) = ((x - 3)^2 - 2)((x - 3)^2 + 1)(x - 2)$; $-c = 2$ is a zero of $f(x)$.

Derivation:

$f(x) = ((x + t)^2 - a)((x + t)^2 - b)(x + c) = ((x + t)^2 - a)((x + t)^2 - b) \cdot (x + c)$
$= ((x + t)^4 - (a + b)(x + t)^2 + ab) \cdot (x + c)$
$= (x(x + t)^4 + c(x + t)^4 - (a + b)x(x + t)^2 - (a + b)c(x + t)^2 + abx + abc)$
$= x^5 + (4t + c)x^4 + \left(6t^2 + 4ct - (a + b)\right)x^3 + (4t^3 + 6ct^2 - 2(a + b)t - (a + b)c)x^2$
$+(t^4 + 4ct^3 - (a + b)t^2 - 2(a + b)ct + ab)x + (ct^4 - (a + b)ct^2 + abc)$
$\Rightarrow$
$k_4 = 4t + c$
$k_3 = 6t^2 + 4ct - (a + b)$
$k_2 = 4t^3 + 6ct^2 - 2(a + b)t - (a + b)c$
$k_1 = t^4 + 4ct^3 - (a + b)t^2 - 2(a + b)ct + ab$
$k_0 = ct^4 - (a + b)ct^2 + abc$

$\Rightarrow$

$c = k_4 - 4t$

$a + b = 6t^2 + 4ct - k_3 = 6t^2 + 4(k_4 - 4t)t - k_3 = -10t^2 + 4k_4t - k_3$

$k_2 = 4t^3 + 6(k_4 - 4t)t^2 - 2(-10t^2 + 4k_4t - k_3)t - (-10t^2 + 4k_4t - k_3)(k_4 - 4t)$

$= -40t^3 + 24k_4t^2 - (4k_4^2 + 2k_3)t + k_4k_3$

$\Leftrightarrow t^3 - \frac{3}{5}k_4t^2 + \left(\frac{1}{10}k_4^2 + \frac{1}{20}k_3\right)t + \frac{1}{40}(k_2 - k_4k_3) = 0$

The equation in t can be resolved, and with t identified, a, b and c can be determined:

$c = k_4 - 4t$

$a + b = -10t^2 + 4k_4t - k_3$

$k_1 = t^4 + 4ct^3 - (a + b)t^2 - 2(a + b)ct + ab$

$= t^4 + 4ct^3 - (-10t^2 + 4k_4t - k_3)t^2 - 2(-10t^2 + 4k_4t - k_3)ct + ab$

$= 11t^4 + 4(6c - k_4)t^3 + (k_3 - 8k_4c)t^2 + 2k_3ct + ab \Leftrightarrow$

$\Rightarrow ab = k_1 - 11t^4 - 4(6c - k_4)t^3 - (k_3 - 8k_4c)t^2 - 2k_3ct$

With $b = -10t^2 + 4k_4t - k_3 - a$

$\Rightarrow a(-10t^2 + 4k_4t - k_3 - a) = k_1 - 11t^4 - 4(6c - k_4)t^3 - (k_3 - 8k_4c)t^2 - 2k_3ct$

$\Leftrightarrow a^2 + (10t^2 - 4k_4t + k_3)a - 11t^4 - 4(6c - k_4)t^3 - (k_3 - 8k_4c)t^2 - 2k_3ct + k_1 = 0$

and a and b can be determined.

4.5 The Case of One Offset Square Root

Looking at the case of one single central square root (cf. chapter 2.6) one may be wondering if a solution approach is possible with a decentralizing offset value. Actually, the case corresponds to the **general, asymmetric case**: each pair of zeroes could serve as the ones with an offset. In such a scenario, the quintic would have 10 possible combinations of pairs of zeroes (with 5 different zeroes; still 6 with one double zero, 3 with two unrelated double zeroes – scenarios which are not covered so far). This indicates already that **a solution does not seem to be possible**, at least in the general case.

The hypothesis is supported by looking at the expressions for the quintic which looks like

$$f(x) := ((x+t)^2 - a)(x-b)(x-c)(x-d)$$
$$= ((x+t)^2 - a)(x^3 - (b+c+d)x^2 + (bc+bd+cd)x - bcd)$$

$$= (x+t)^2x^3 - (b+c+d)(x+t)^2x^2 + (bc+bd+cd)(x+t)^2x - bcd(x+t)^2$$
$$-ax^3 + a(b+c+d)x^2 - a(bc+bd+cd)x + abcd$$

$$= x^5 + (2t-(b+c+d))x^4 + (t^2 - 2t(b+c+d) + (bc+bd+cd) - a)x^3$$
$$+\big((a-t^2)(b+c+d) + 2t(bc+bd+cd) - bcd\big)x^2$$
$$+(-(a-t^2)(bc+bd+cd) - 2tbcd)x + (a-t^2)bcd$$

It is clearly visible that linear and quadratic expressions of t including their combinations occur in every coefficient of $f(x)$ (except in k_4 where it occurs as linear expression only). Together with the variables a, b, c, d, the coefficients represent a non-linear set of five equations with quadratic expressions and variables coupled as factors. One can imagine that this cannot be resolved because a resolution approach would end up in another polynomial equation of degree five or higher, a rational equation and/or in one with two variables which depend on each other and cannot be generally solved. For example, it can be derived that

$$k_4 = 2t - (b+c+d) \Leftrightarrow \boldsymbol{b+c+d} = 2t - k_4$$

$$k_3 = t^2 - 2t(b+c+d) + (bc+bd+cd) - a = t^2 - 2t\left(2t - k_4\right) + (bc+bd+cd) - a$$
$$= -3t^2 + 2k_4t + (bc+bd+cd) - a \Leftrightarrow \boldsymbol{bc+bd+cd} = 3t^2 - 2k_4t + k_3 + a$$

$$k_2 = (a-t^2)(b+c+d) + 2t(bc+bd+cd) - bcd$$
$$= (a-t^2)\left(2t - k_4\right) + 2t\left(3t^2 - 2k_4t + k_3 + a\right) - bcd$$
$$-2t^3 + k_4t^2 + 2at - k_4a + 6t^3 - 4k_4t^2 + 2(k_3+a)t - bcd$$
$$= 4t^3 - 3k_4t^2 + 2(k_3+2a)t - k_4a - bcd \Leftrightarrow \boldsymbol{bcd} = 4t^3 - 3k_4t^2 + 2(k_3+2a)t - k_4a - k_2$$

$$k_1 = -(a-t^2)(bc+bd+cd) - 2tbcd$$
$$= -(a-t^2)\left(3t^2 - 2k_4t + k_3 + a\right) - 2t(4t^3 - 3k_4t^2 + 2(k_3+2a)t - k_4a - k_2)$$
$$= 3t^4 - 2k_4t^3 + (k_3+a)t^2 - 3at^2 + 2k_4at - a(k_3+a) - 8t^4 + 6k_4t^3 - 4(k_3+2a)t^2 + 2(k_4a+k_2)t$$
$$= -5t^4 + 4k_4t^3 + (-3k_3 - 10a)t^2 + (4k_4a + 2k_2)t - a(k_3+a)$$

$$k_0 = (a-t^2)bcd = (a-t^2)(4t^3 - 3k_4t^2 + 2(k_3+2a)t - k_4a - k_2)$$
$$= -4t^5 + 3k_4t^4 - 2k_3t^3 + (k_2 - 2k_4a)t^2 + 2a(k_3+2a)t - a(k_4a+k_2)$$

Nevertheless, it should be mentioned that the relationship between t and a actually can be resolved, e.g. in the way that a can be expressed by t (using a combination of both equations for k_0 and k_1 to get the 5th power of t eliminated and all available information combined):

$$k_0 - \frac{4}{5}tk_1 = -\frac{1}{5}k_4t^4 + \left(\frac{2}{5}k_3 + 8a\right)t^3 + \left(k_2 - 2k_4a - \left(\frac{16}{5}k_4a + \frac{8}{5}k_2\right)\right)t^2$$

$$+\left(2a(k_3 + 2a) + \frac{4}{5}a(k_3 + a)\right)t - a(k_4a + k_2)$$

$$\Leftrightarrow (\frac{24}{5}t - k_4)a^2 + \left(8t^3 - \frac{26}{5}k_4t^2 + \frac{14}{5}k_3t - k_2\right)a - \frac{1}{5}k_4t^4 + \frac{2}{5}k_3t^3 - \frac{3}{5}k_2t^2 + \frac{4}{5}k_1t - k_0 = 0$$

If $t = \frac{5}{24}k_4$, the equation is linear and, under the condition that the denominator is $\neq 0$,

$$a = \frac{\frac{1}{5}k_4t^4 - \frac{2}{5}k_3t^3 + \frac{3}{5}k_2t^2 - \frac{4}{5}k_1t + k_0}{8t^3 - \frac{26}{5}k_4t^2 + \frac{14}{5}k_3t - k_2}$$

If $t \neq \frac{5}{24}k_4$, the equation is quadratic and a can be determined by solving the equation

$$\Leftrightarrow a^2 + \frac{8t^3 - \frac{26}{5}k_4t^2 + \frac{14}{5}k_3t - k_2}{\frac{24}{5}t - k_4}a + \frac{-\frac{1}{5}k_4t^4 + \frac{2}{5}k_3t^3 - \frac{3}{5}k_2t^2 + \frac{4}{5}k_1t - k_0}{\frac{24}{5}t - k_4} = 0$$

$$\Rightarrow a_{1,2} = -\frac{4t^3 - \frac{13}{5}k_4t^2 + \frac{7}{5}k_3t - \frac{1}{2}k_2}{\frac{24}{5}t - k_4}$$

$$\pm\sqrt{\left(\frac{4t^3 - \frac{13}{5}k_4t^2 + \frac{7}{5}k_3t - \frac{1}{2}k_2}{\frac{24}{5}t - k_4}\right)^2 - \frac{-\frac{1}{5}k_4t^4 + \frac{2}{5}k_3t^3 - \frac{3}{5}k_2t^2 + \frac{4}{5}k_1t - k_0}{\frac{24}{5}t - k_4}}$$

For each pair of zeroes of $f(x)$, a (the proper one of a_1 or a_2) could be called a *discriminant* of $f(x)$. If $a = 0$, $f(x)$ has a double zero at $-t$. In case a symmetry value t between two zeroes happens to be known, the related one or two zeroes can be determined by one of the a values as

$$n_{1,2} = \pm\sqrt{a} - t$$

Example:
$f(x) = x^5 + x^4 - 7x^3 - 33x^2 - 66x - 40$
Assume we know $t = 1$. Then we can calculate a *and* $\pm\sqrt{a}$ with the quadratic equation:

$$a = -\frac{P}{2} \pm \sqrt{\frac{P2}{4} - Q} \text{ with}$$

$$P := 2 \cdot \frac{4 - \frac{13}{5} + \frac{7}{5}\cdot(-7) - \frac{1}{2}\cdot(-33)}{\frac{24}{5} - 1} = \frac{81}{19}$$

$$Q := \frac{-\frac{1}{5} + \frac{2}{5}\cdot(-7) - \frac{3}{5}\cdot(-33) + \frac{4}{5}\cdot(-66) + 40}{\frac{24}{5} - 1} = \frac{5}{19} \cdot \frac{-15 + 99 - 264 + 200}{5} = \frac{20}{19}$$

$$a = -\frac{81}{2\cdot19} \pm \sqrt{\frac{81^2}{2^2\cdot19^2} - \frac{1520}{2^2\cdot19^2}} = -\frac{81 \pm 71}{2\cdot19}$$

For $a = -\frac{81+71}{2\cdot 19} = -4$ the roots are $\pm\sqrt{a} = \pm 2i$, thus with that, two zeroes of $f(x)$ are $-1 \pm 2i$.
Calculating a and $\pm\sqrt{a}$ for $t = -\frac{3}{2}$ accordingly:

$$P := 2 \cdot \frac{4\cdot\left(-\frac{27}{8}\right) - \frac{13}{5}\cdot\frac{9}{4} + \frac{7}{5}\cdot(-7)\left(-\frac{3}{2}\right) - \frac{1}{2}\cdot(-33)}{\frac{24}{5}\cdot\left(-\frac{3}{2}\right)-1} = -\frac{237}{2\cdot 41}$$

$$Q := \frac{-\frac{1}{5}\cdot\frac{81}{16} + \frac{2}{5}\cdot(-7)\left(-\frac{27}{8}\right) - \frac{3}{5}\cdot(-33)\cdot\frac{9}{4} + \frac{4}{5}\cdot(-66)\left(-\frac{3}{2}\right) + 40}{\frac{24}{5}\cdot\left(-\frac{3}{2}\right)-1} = \frac{\frac{81}{16} - \frac{189}{4} - \frac{891}{4} - 396 - 200}{41} = \frac{81}{4^2\cdot 41} - \frac{866}{41} = -\frac{564775}{4^2\cdot 41^2}$$

$$a = \frac{237}{4\cdot 41} \pm \sqrt{\frac{237^2+564775}{4^2\cdot 41^2}} = \frac{237}{4\cdot 41} \pm \frac{788}{4\cdot 41} \Rightarrow a = \frac{25}{4} \vee a = -\frac{551}{4\cdot 41}$$

For $a = \frac{25}{4}$ the roots $\pm\sqrt{a} = \pm\frac{5}{2}$ are found, thus two zeroes of $f(x)$ are $\frac{3}{2} \pm \frac{5}{2}$, i.e. -1 and 4.

In an analogous way, it can be found that $t = -1$ and $t = +\frac{3}{2}$ also lead to solutions, with -2 as the remaining zero

$$\Rightarrow f(x) = (x + 1 + 2i)(x + 1 - 2i)(x + 2)(x + 1)(x - 4)$$

Example:
$f(x) = x^5 - 12x^4 + 49x^3 - 66x^2 - 32x + 96$
Assume we know $t = -\frac{5}{2}$. Then, the condition $t = \frac{5}{24}k_4$ is true, and we can calculate a *and* $\pm\sqrt{a}$ with the linear equation:

$$a = \frac{\frac{1}{5}(-12)\frac{625}{16} - \frac{2}{5}49\left(-\frac{125}{8}\right) + \frac{3}{5}(-66)\frac{25}{4} - \frac{4}{5}(-32)\left(-\frac{5}{2}\right) + 96}{8\left(-\frac{125}{8}\right) - \frac{26}{5}(-12)\frac{25}{4} + \frac{14}{5}49\left(-\frac{5}{2}\right) - (-66)} = \frac{-\frac{3\cdot 125}{4} + \frac{49\cdot 25}{4} - \frac{3\cdot 66\cdot 5}{4} - \frac{4\cdot 32\cdot 2}{4} + \frac{96\cdot 4}{4}}{-125 + 26\cdot 3\cdot 5 - 7\cdot 49 + 66} = \frac{-12}{4\cdot(-12)} = \frac{1}{4}$$

Thus, $f(x) = \left(x - \frac{5}{2} - \frac{1}{2}\right)\left(x - \frac{5}{2} + \frac{1}{2}\right) g_3(x) = (x-3)(x-2)(x-4)^2(x+1)$

Additional conclusion: if $f(x) \in \mathbb{R}[X]$, at least one $t \in \mathbb{R}$ does exist.

This can be derived by the following consideration:
if $f(x) \in \mathbb{R}[X]$, there is at least one $n_0 \in \mathbb{R}$ with $f(n_0) = 0$. If another $n_1 \in \mathbb{R}$ with $f(n_1) = 0$, a value for $t \in \mathbb{R}$ has been found with $t = \frac{n_0+n_1}{2}$. Otherwise, $g_4(x) = f(x){:}\,(x - n_0)$ is a quartic polynomial which has two pairs of complex conjugated zeroes [7]. If the zeroes in one of these pairs are named n_1 and n_2, then $t \in \mathbb{R}$ has been found with $t = \frac{n_1+n_2}{2} = Re(n_1) = Re(n_2)$.

One further significant simplification actually occurs if t itself is a single zero as well. This happens for example if there are three zeroes n_1, n_2, n_3 on a straight line, e.g. $\in \mathbb{R}$, with $t := -n_2$ and a constant distance $|\sqrt{a}|$ between n_1, n_2 and n_2, n_3.

If $f(t) = 0$, we for example can set $d := t$ and get to the following much simpler expressions:

$b + c + d = 2t - k_4 \Leftrightarrow b + c = 3t - k_4$

$bc + bd + cd = 3t^2 - 2k_4t + k_3 + a \Leftrightarrow bc = 6t^2 - 3k_4t + k_3 + a$

$bcd = 4t^3 - 3k_4t^2 + 2(k_3 + 2a)t - k_4a - k_2 \wedge bcd = (6t^2 - 3k_4t + k_3 + a)\cdot(-t)$

$$\Rightarrow 10t^3 - 6k_4t^2 + (3k_3 + 5a)t - (k_4a + k_2) = 0$$

The last expression also can be written as

$$10t^3 - 6k_4t^2 + 3k_3t - k_2 = a \cdot (k_4 - 5t)$$

in other words, the left side polynomial of degree 3 is set equal to a multiple (by a) of the linear one on the right side of the equation. If $k_4 \neq 5t$, this results in

$$a = \frac{10t^3 - 6k_4t^2 + 3k_3t - k_2}{k_4 - 5t}$$

In the special case that $k_4 = 5t$, one zero $-t$ has already been found, and the others can be determined by polynomial division of $f(x): (x + t)$.

Example:
$f(x) = x^5 - 2x^4 - 12x^3 + 32x^2 - 13x - 6$
Testing $t = -2$ with $n_2 = -t = 2$ as a zero:

$a = \frac{-80+48+72-32}{-2+10} = \frac{8}{8} = 1 \Rightarrow f(x) = (x-1)(x-2)(x-3) \cdot g_2(x)$
$g_2(x)$ can be determined as $g_2(x) = x^2 + 4x + 1$ by polynomial division
$g_2(x) = f(x): ((x-1)(x-2)(x-3))$

5 The Algorithm and More Examples

In the previous chapters, "nice" quintics were chosen to demonstrate how to derive and calculate zeroes in different ways. In general, it is not obvious by the coefficients which scenario is given. Thus, an algorithm could be applied which checks if the conditions of the different scenarios are fulfilled, in order of increasing complexity for efficiency. By this approach, the zeroes can be determined if they are located in $\mathbb{C}$ in any of the described circular symmetries.

In the following table, a few more examples are shown and how to get them resolved by the matching method as described above.

4th root scenario (chapter 2.2; [1])	$\boldsymbol{f(x) = x^5 - x^4 - x + 1}$ f fulfils the conditions $k_3 = k_2 = 0; k_4 \neq 0 \wedge k_1 \neq 0; k_0 = k_1 \cdot k_4$ Four of the roots are $n_k = re^{\frac{2\pi ik}{4}}; k = 0, \ldots, 3; r = 1$ For example, $n_0 = \sqrt[4]{1} = 1; n_1 = e^{\frac{2\pi i}{4}} = i$ Thus, $f(x) = (x-1)(x+1)(x-i)(x+i)(x-b)$ $= (x^2-1)(x^2+1)(x-b) = (x^4-1)(x-b)$ The 5th zero is $b = 1$.
3rd root scenario (chapter 2.4)	$\boldsymbol{f(x) = x^5 - 2x^4 + 3x^3 + 2x^2 - 4x + 6}$ f fulfils the conditions $k_i \neq 0 \wedge k_1 = k_2 \cdot k_4 \wedge k_0 = k_2 \cdot k_3$ With $a = -k_2 = 2 \Rightarrow b = 1 + \sqrt{2}i \wedge c = 1 - \sqrt{2}i$ Thus, $f(x) = (x^3 - a)(x-b)(x-c) =$ $(x^3+2)\left(x - 1 - \sqrt{2}i\right)\left(x - 1 + \sqrt{2}i\right) = (x^3+2)(x^2 - 2x + 3)$
2nd root scenario (chapter 2.6)	$\boldsymbol{f(x) = x^5 + 2x^4 - x^3 + 2x^2 - 20x - 24}$ f fulfils the conditions of that scenario. With this assumption in the beginning $\Rightarrow a = -\frac{1}{2} \pm \sqrt{\frac{1}{4} + \frac{24}{2}} \Rightarrow a = -4 \vee a = 3$ With $a = -4 \Rightarrow f(x): (x^2+4) = x^3 + 2x^2 - 5x - 6$ $= (x-2)(x-1)(x+3)$ $\Rightarrow \pm\sqrt{a} = \pm 2i \wedge b = 2; c = 1; d = -3$ Thus, $f(x) = (x+2i)(x-2i)(x-2)(x-1)(x+3)$
2nd root scenario (chapter 2.6)	$\boldsymbol{f(x) = x^5 + \sqrt{5}ix^4 + 5x^3 + 9\sqrt{5}ix^2 + \left(24 + 6\sqrt{5}i\right)x + \left(10 + 40\sqrt{5}i\right)}$ f fulfils the conditions of that scenario. With this assumption in the beginning $\Rightarrow a = -\frac{9}{2} \pm \sqrt{\frac{81}{4} - \frac{10+40\sqrt{5}i}{\sqrt{5}i}} = -\frac{9}{2} \pm \sqrt{-\frac{79}{4} + 2\sqrt{5}i} = -\frac{9}{2} \pm (\frac{1}{2} + 2\sqrt{5}i)$ With $a = -4 + 2\sqrt{5}i \Rightarrow \pm\sqrt{a} = \pm(1 + \sqrt{5}i) \Rightarrow$ $f(x) = \left(x + 1 + \sqrt{5}i\right)\left(x - 1 - \sqrt{5}i\right)(x + 1 - 2i)(x + 1 + 2i)\left(x - 2 + \sqrt{5}i\right)$ after polynomial division $f(x): (x^2 + 4 - 2\sqrt{5}i)$

3-times zero scenario (chapter 3.3)	$\boldsymbol{f(x) = x^5 - \frac{20}{3}x^3 + 20x - \frac{32}{3}\sqrt{2}}$ f fulfils the conditions of that scenario. With this assumption in the beginning and scenario $k_4 = 0$, a fulfils the equation $a^3 - \frac{3}{4} \cdot 0a^2 + \frac{5}{4} \cdot \left(\frac{-32}{3}\sqrt{2}\right) \cdot \left(-\frac{3}{20}\right) = a^3 + 2\sqrt{2} = 0$ The solutions $a_{1,2,3}$ of this equation are the respective 3rd roots with $\lvert a_i \rvert = \sqrt{2}$ (cf. chapter 2.4). With $a = -\sqrt{2}$, the quintic can be resolved and b, c determined: $b, c = \frac{0+3\sqrt{2}}{2} \pm \sqrt{\frac{18}{4} - \frac{16}{3}} = \frac{3\sqrt{2}}{2} \pm \sqrt{\frac{5}{6}}i$ $\Rightarrow f(x) = \left(x - \sqrt{2}\right)^3 \left(x + \frac{3\sqrt{2}}{2} + \sqrt{\frac{5}{6}}i\right)\left(x + \frac{3\sqrt{2}}{2} - \sqrt{\frac{5}{6}}i\right)$
3-times zero scenario (chapter 3.3)	$\boldsymbol{f(x) = x^5 + \frac{22}{3}x^4 - 4x^3 - 144x^2 + \frac{392\sqrt{7} - 20}{3}x - 216}$ f fulfils the conditions of that scenario. With this assumption in the beginning and scenario $k_4 \neq 0$, a fulfils the equation $a^4 - \frac{4}{3} \cdot \frac{(-4)\cdot 3}{22}a^3 + \frac{(-144)\cdot 3}{22}a^2 - \frac{5}{3} \cdot \frac{(-216)\cdot 3}{22} = a^4 + \frac{8}{11}a^3 - \frac{216}{11}a^2 + \frac{540}{11} = 0$ A solution can be found by linear Tschirnhaus transformation and solving the cubic resolvent as described. [8] Transformation of the equation in a to y in order to eliminate the a^3 expression with $k := \frac{8}{11\cdot 4} = \frac{2}{11}$ and $y := a + k \Leftrightarrow a = y - k$: $(y-k)^4 + \frac{8}{11}(y-k)^3 - \frac{216}{11}(y-k)^2 + \frac{540}{11} = 0$ $\Leftrightarrow y^4 + 0 \cdot y^3 - \frac{2400}{11^2}y^2 + \frac{9568}{11^3}y + \frac{709188}{11^4} = 0$ The next step is solving the resolvent equation $\prod_{i=1}^{3}(t - t_i) = t^3 + \frac{2400}{11^2}t^2 - 4 \cdot \frac{709188}{11^4}t - 4 \cdot \frac{2400}{11^2} \cdot \frac{709188}{11^4} - \frac{9568^2}{11^6}$ The arithmetic becomes rather effortful; the cubic equation can be solved by the Cardano formulas. Once the three t_i are found, the zeroes y_i of the equation in y of degree 4 above can be derived by the resolvent equations in t_i: $t_1 = y_1y_2 + y_3y_4; t_2 = y_1y_3 + y_2y_4; t_3 = y_1y_4 + y_2y_3; \sum_{i=1}^{4} y_i = 0$ Finally, the y_i need to be transformed back to the a_i by $a_i = y_i - k$. Selecting the proper a_i provides the three times zero a of the quintic. The transformations and calculations are not further pursued here – they should be executed by computational algorithms. Since the example has been constructed, the zero a is known as $a = 1 - \sqrt{7}$ $\Rightarrow b, c = \frac{\frac{22}{3} - 3(1-\sqrt{7})}{2} \pm \sqrt{\frac{\left(\frac{22}{3} - 3(1-\sqrt{7})\right)^2}{4} - \frac{-216}{(1-\sqrt{7})^3}}$ $= \frac{13+9\sqrt{7}}{6} \pm \sqrt{\frac{368+117\sqrt{7}}{18} - \frac{-216}{22-10\sqrt{7}}}$ $= \frac{13+9\sqrt{7}}{6} \pm \sqrt{\frac{368+117\sqrt{7}}{18} - \left(22 + 10\sqrt{7}\right)} = \frac{13+9\sqrt{7}}{6} \pm \sqrt{-\frac{14}{9} - \frac{7}{2}\sqrt{7}}$ $\Rightarrow f(x) = \left(x + 1 - \sqrt{7}\right)^3 \cdot$ $\left(x + \frac{13+9\sqrt{7}}{6} + \sqrt{-\frac{14}{9} - \frac{7}{2}\sqrt{7}}\right)\left(x + \frac{13+9\sqrt{7}}{6} - \sqrt{-\frac{14}{9} - \frac{7}{2}\sqrt{7}}\right)$

4th offset root scenario (chapter 4.2)	$\boldsymbol{f(x) = x^5 + 2x^4 - \frac{13}{2}x^3 + \frac{11}{2}x^2 - \frac{15}{16}x + \frac{17}{4}}$ *f* fulfils the conditions of that scenario. With this assumption in the beginning, *t* is calculated as $t = \frac{2}{5} \pm \sqrt{\frac{4}{25} + \frac{13}{20}} \Rightarrow t = \frac{13}{10} \vee t = -\frac{1}{2}$ Choosing $t = -\frac{1}{2} \Rightarrow b = 2 - 4 \cdot \left(-\frac{1}{2}\right) = 4 \Rightarrow a = \left(-\frac{1}{2}\right)^4 - \frac{17}{4 \cdot 4} = -1$ The 4th roots of -1 are $\pm\frac{1}{\sqrt{2}}(1 \pm i)$ $f(x) = \left(x - \frac{1}{2} + \frac{1}{\sqrt{2}}(1+i)\right)\left(x - \frac{1}{2} - \frac{1}{\sqrt{2}}(1+i)\right) \cdot$ $\left(x - \frac{1}{2} + \frac{1}{\sqrt{2}}(1-i)\right)\left(x - \frac{1}{2} - \frac{1}{\sqrt{2}}(1-i)\right) \cdot (x+4)$
Two offset square roots scenario (chapter 4.4)	$\boldsymbol{f(x) = x^5 + \frac{6}{5}x^4 - \frac{51}{10}x^3 - \frac{17}{10}x^2 + \frac{673}{80}x - \frac{77}{20}}$ *f* fulfils the conditions of that scenario. With this assumption in the beginning, *t* fulfils the equation $t^3 - \frac{3}{5} \cdot \frac{6}{5}t^2 + \left(\frac{36}{25 \cdot 10} + \frac{-51}{20 \cdot 10}\right)t + \frac{1}{40}\left(-\frac{17}{10} + \frac{306}{50}\right)$ $= t^3 - \frac{18}{25}t^2 - \frac{111}{1000}t + \frac{221}{2000} = 0$ A solution can be found by linear Tschirnhaus transformation (see example above) and solving the cubic equation. With the result $t = \frac{1}{2}$, zeroes can be calculated: $c = \frac{6}{5} - \frac{4}{2} = -\frac{4}{5}$ and $a^2 + \left(\frac{5}{2} - \frac{12}{5} - \frac{51}{10}\right)a - \frac{11}{16} - 4\left(-6 \cdot \frac{4}{5} - \frac{6}{5}\right)\frac{1}{8} - \left(-\frac{51}{10} + \frac{192}{25}\right)\frac{1}{4} - \frac{102}{25} + \frac{673}{80}$ $= a^2 - 5a - \frac{11}{16} + 3 - \frac{129}{200} - \frac{102}{25} + \frac{673}{80} = a^2 - 5a + 6 = 0$ $\Rightarrow a = \frac{5}{2} \pm \sqrt{\frac{25}{4} - 6} = \frac{5}{2} \pm \frac{1}{2}$ $\Rightarrow \pm\sqrt{a} = \pm\sqrt{2} \vee \pm\sqrt{a} = \pm\sqrt{3}$ $\Rightarrow f(x) = (x - \frac{4}{5})(x + \frac{1}{2} + \sqrt{2})(x + \frac{1}{2} - \sqrt{2})(x + \frac{1}{2} + \sqrt{3})(x + \frac{1}{2} - \sqrt{3})$
Two offset square roots scenario (chapter 4.4)	$\boldsymbol{f(x) = x^5 - 4\sqrt{2}x^4 + 12x^3 - 10\sqrt{2}x^2 + \frac{35}{4}x - \frac{3}{2}\sqrt{2}}$ *f* fulfils the conditions of that scenario. With this assumption in the beginning, *t* fulfils the equation $t^3 - \frac{3}{5} \cdot \left(-4\sqrt{2}\right)t^2 + \left(\frac{32}{10} + \frac{12}{20}\right)t + \frac{1}{40}\left(-10\sqrt{2} + 48\sqrt{2}\right)$ $= t^3 + \frac{12}{5}\sqrt{2}t^2 + \frac{19}{5}t + \frac{19}{20}\sqrt{2} = 0$ As above, a solution can be found by linear Tschirnhaus transformation and solving the cubic equation. With *SageMath*, three solutions for *t* with a rather complicated structure are found. One of them is the right value for t, which is $t = \frac{1}{10 \cdot \sqrt[3]{9}} \cdot \sqrt[3]{25\sqrt{38}\sqrt{3} + 189\sqrt{2}} - \frac{4}{5}\sqrt{2} + \frac{6}{5 \cdot \sqrt[3]{9}^2 \cdot \sqrt[3]{25\sqrt{38}\sqrt{3} + 189\sqrt{2}}}$

	Based on that, the quadratic equation in a leads to a solution for a, b, c which represent zeroes of $f(x)$ by $\pm\sqrt{a} - t, \pm\sqrt{b} - t, -c$. However, the found expressions are really complicated – too large to be printed here. Nevertheless, it could be shown with *SageMath* that these are indeed zeroes of $f(x)$.
Two offset square roots scenario (chapter 4.4)	$\boldsymbol{f(x) = x^5 - 12x^4 + 120x^3 - (650 + 310i)x^2 + (979 + 1800i)x + (442 - 2210i)}$ f fulfils the conditions of that scenario. With this assumption in the beginning, t fulfils the equation $t^3 + \frac{3}{5} \cdot 12t^2 + \left(\frac{144}{10} + \frac{120}{20}\right)t + \frac{-(650+310i) + 1440}{40}$ $= t^3 + \frac{36}{5}t^2 + \frac{102}{5}t + \frac{79 - 31i}{4} = 0$ As above, a solution can be found by linear Tschirnhaus transformation and resolving the cubic equation. *SageMath* outputs three solutions for t with a rather complicated structure; one of them is $t = -\frac{1}{20}(i\sqrt{3} + 1) \cdot \sqrt[3]{5\sqrt{242110i - 531232} + 3875i + 781}$ $+ \frac{26 \cdot (-i\sqrt{3}+1)}{5 \cdot \sqrt[3]{5\sqrt{242110i - 531232} + 3875i + 781}} - \frac{12}{5}$ In fact, this is nothing else than $t = -\frac{5}{2}(1 + i)$ as can be seen with $t^2 = \frac{25}{2}i;\ t^3 = \frac{125}{4}(1 - i)$ by $\frac{125}{4}(1 - i) + \frac{36}{5} \cdot \frac{25}{2}i - \frac{102}{5} \cdot \frac{5}{2}(1 + i) + \frac{79 - 31i}{4}$ $= \left(\frac{125}{4} - \frac{204}{4} + \frac{79}{4}\right) + \left(-\frac{125}{4} + \frac{360}{4} - \frac{204}{4} - \frac{31}{4}\right)i = 0$ Similarly like in the examples above, with t some zeroes can be calculated: $a = -\frac{5}{2}i \pm \sqrt{-\frac{25}{4} + \frac{9}{4}} = -\frac{5}{2}i \pm 2i \Rightarrow a = -\frac{1}{2}i \vee a = -\frac{9}{2}i$ The result reveals four zeroes: for $a = -\frac{1}{2}i \Rightarrow \pm\sqrt{a} = \pm\frac{1}{2}(i - 1)$ $\Rightarrow f(t + \sqrt{a}) = f(2 + 3i) = 0 = f(t - \sqrt{a}) = f(3 + 2i)$ for $a = -\frac{9}{2}i \Rightarrow \pm\sqrt{a} = \pm\frac{3}{2}(i - 1)$ $\Rightarrow f(t + \sqrt{a}) = f(1 + 4i) = 0 = f(t - \sqrt{a}) = f(4 + i)$ $\Rightarrow f(x) = (x - 2 - 3i)(x - 3 - 2i)(x - 1 - 4i)(x - 4 - i)(x - 2 + 10i)$

6 Additional Clues

6.1 Sum of Squares of Zeroes

In addition, it is possible to derive another rather simple equation between the zeroes and two coefficients of the quintic.

Starting with $f(x) = \sum k_i x^i = (x+a)(x+b)(x+c)(x+d)(x+e)$, the following is true for k_3 and k_4 and the zeroes $-a, \ldots, -e$ of f. Here, let $-e$ be an arbitrarily selected zero.

$$\begin{aligned}
k_4 &= a+b+c+d+e \Leftrightarrow a+b+c+d = k_4 - e \\
k_3 &= ab+ac+ad+ae+bc+bd+be+cd+ce+de \\
&= (ab+cd)+(ac+bd)+(ad+bc)+(a+b+c+d)\cdot e \\
&= (ab+cd)+(ac+bd)+(ad+bc)+k_4e-e^2
\end{aligned}$$

The relationships shown above are true for all zeroes of f. Each zero could play the role of e; thus, by summing up across all zeroes in the role of e, the equation can be brought into a simple, symmetric form as sum of squares of all zeroes.

$$\begin{aligned}
&a^2+b^2+c^2+d^2+e^2-k_4(a+b+c+d+e)+5k_3 \\
&-\big((ab+cd)+(ac+bd)+(ad+bc)\big) \\
&-\big((ab+ce)+(ac+be)+(ae+bc)\big) \\
&-\big((ab+de)+(ad+be)+(ae+bd)\big) \\
&-\big((ac+de)+(ad+ce)+(ae+cd)\big) \\
&-\big((bc+de)+(bd+ce)+(be+cd)\big) = 0
\end{aligned}$$

$$\begin{aligned}
&\Leftrightarrow a^2+b^2+c^2+d^2+e^2-k_4^2+5k_3 \\
&-3\cdot(ab+ac+ad+ae+bc+bd+be+cd+ce+de) \\
&= a^2+b^2+c^2+d^2+e^2-k_4^2+5k_3-3k_3 = 0
\end{aligned}$$

$$\Leftrightarrow \boldsymbol{a^2+b^2+c^2+d^2+e^2 = k_4^2-2k_3}$$

That this condition indeed holds is demonstrated by using the following examples.

$f(x) = x^5 - 3x^4 - \frac{64}{9}x^3 + \frac{100}{9}x^2 + \frac{64}{9}x - \frac{64}{9} = (x-\frac{4}{3})(x-\frac{2}{3})(x^3-x^2-10x-8)$
$= (x-\frac{4}{3})(x-\frac{2}{3})(x+1)(x+2)(x-4)$. The above equation looks like:

$$\left(\frac{2}{3}\right)^2 + \left(\frac{4}{3}\right)^2 + (-1)^2 + (-2)^2 + 4^2 = \frac{4}{9} + \frac{16}{9} + 21 = 23\frac{2}{9}$$
$$= (-3)^2 - (2\cdot(-\frac{64}{9})) = 9 + \frac{128}{9} = 23\frac{2}{9}$$

$f(x) = x^5 + \sqrt{5}ix^4 + 5x^3 + 9\sqrt{5}ix^2 + \big(24+6\sqrt{5}i\big)x + \big(10+40\sqrt{5}i\big)$
$k_4^2 - 2k_3 = -5 - 10 = -15 \in \mathbb{R};\ k_4 = \sqrt{5}i \in Im(\mathbb{C}) \wedge k_3 \in \mathbb{R}$.
The quintic can be resolved as shown above; the sum of squares of zeroes is
$2\cdot\big(-4+2\sqrt{5}i\big) + (-3-4i) + (-3+4i) + \big(-1-4\sqrt{5}i\big) = -15$

A consequence of that equation is that all zeroes of a quintic are $\in \mathbb{R}$ only if $k_4^2 - 2k_3 \in \mathbb{R}$ and > 0 (necessary condition). In addition: for each zero $n \Rightarrow |n| < \sqrt{k_4^2 - 2k_3}$ in that case. In the first example above, the absolute value of each zero n can be estimated by

$$|n| < \sqrt{9 + \frac{128}{9}} = \frac{\sqrt{209}}{3} < 4.82$$

6.2 The Constant Value

Since $k_0 = -abcde$ with a, b, c, d, e the zeroes of $f(x)$, the absolute value can be quantified as $|k_0| = |abcde| = |a||b||c||d||e|$

It can be safely stated that there must be a zero (named a here) fulfilling the condition

$$|a| \leq |\sqrt[5]{k_0}| = \sqrt[5]{|k_0|}$$

since if one zero has a higher absolute value, there must be another one with a lower value. Furthermore, $|k_0|$ itself reveals clues on potential solutions.

> Example (see also chapter 5, last example):
> $f(x) = x^5 - 12x^4 + 120x^3 - (650 + 310i)x^2 + (979 + 1800i)x + (442 - 2210i)$
>
> $|k_0| = \sqrt{442^2 + 2210^2} = \sqrt{5079464} = \sqrt{2} \cdot \sqrt{13} \cdot 2 \cdot 13 \cdot 17$
> $\sqrt[5]{|k_0|} = \sqrt[5]{\sqrt{2} \cdot \sqrt{13} \cdot 2 \cdot 13 \cdot 17} < 4.7$
> $\Rightarrow$ one may now look at potential zeros with absolute value smaller than that value. Furthermore, the prime factor decomposition indicates some obvious options - being aware that there is an infinite number of possibilities - but together with the values visible in k_4, k_3 and C, possibilities potentially could be narrowed down.
> With e.g. 13 as a prime factor of $|k_0|$, the numbers $a = \pm(2 \pm 3i)$ and $= \pm(3 \pm 2i)$ with $|a| = \sqrt{13}$ could be checked (or similarly, other combinations of rational real and imaginary parts). Polynomial division of $f(x)$ by $(2 + 3i)$ or by $(3 + 2i)$ confirms a solution:
>
> $f(x){:}\,(x - (2 + 3i)) = x^4 + (-10 + 2i)x^3 + (84 - 16i))x^2 - (354 + 90i)x + 391 + 408i$
>
> Similarly, it works for $(3 + 2i)$. With that, one can execute polynomial division further and identify the remaining zeroes (see above).

If $k_0 \in \mathbb{R}$, there is another rather strong condition since $k_0 = -abcde$ for the zeros a, b, c, d, e. For this case, an interesting equation can be derived (with $a_r, a_i, b_r, b_i, c_r, c_i, d_r, d_i, e_r, e_i \in \mathbb{R}$):

$$k_0 = -(a_r + a_i i)(b_r + b_i i)(c_r + c_i i)(d_r + d_i i)(e_r + e_i i)$$

$$= -(a_r b_r - a_i b_i + (a_r b_i + a_i b_r)i)(c_r d_r - c_i d_i + (c_r d_i + c_i d_r)i)(e_r + e_i i)$$

Now define

$$R := (a_r b_r - a_i b_i)(c_r d_r - c_i d_i) - (a_r b_i + a_i b_r)(c_r d_i + c_i d_r)$$

$$I := (a_r b_r - a_i b_i)(c_r d_i + c_i d_r) + (a_r b_i + a_i b_r)(c_r d_r - c_i d_i)$$

$$\Rightarrow k_0 = -(R + Ii)(e_r + e_i i) = -(Re_r - Ie_i + (Re_i + Ie_r)i)$$

If $k_0 \in \mathbb{R} \Rightarrow Re_i + Ie_r = 0$

This condition must be symmetrically met for all zeroes (in the role of $e_r + e_i i$).

6.3 Transcendental Zeroes

It seems to be obvious - a supposition, not proven here - that with transcendental zeroes, not all coefficients can be algebraic - even if some can be.

An example how a transcendental zero propagates into coefficients: assume among the zeros are sin(1) and cos(1), e.g.

$f(x) = (x+1)(x+\cos(1))(x-\cos(1))(x+\sin(1))(x-\sin(1))$

$= (x+1)(x^2 - cos^2(1))(x^2 - sin^2(1))$

$=(x+1)(x^4 - x^2 + cos^2(1)\, sin^2(1))$

$= x^5 + x^4 - x^3 - x^2 + \frac{sin^2(2)}{4}\, x + \frac{sin^2(2)}{4}$

Transcendental values only in a few coefficients indicate that some analytical theorems may apply, like for instance the addition and multiplication theorems of the trigonometric functions *cos* and *sin* in the example above.

7 Outlook

It has been shown that, for a considerable range of quintics - those which have a circular symmetry of zeroes -, zeroes actually can be determined by resolvents which are derived in this book. Further research could engage in:

- Finding additional (potentially specific) cases of the general (asymmetric) scenario which can be solved. For example, the equation with the sum of squares of zeroes, in combination with other identities, could help to get insight into additional cases.
- The analysis of algebraic vs. transcendental values for trigonometric functions - more generally the exponential function in $\mathbb{C}$ - could lead to further solutions.
- Developing algorithms which can be implemented in computer software to automate calculations and checks of the key equations, supported by polynomial division and multiplication in $\mathbb{C}[X]$ and calculating exponential expressions and radicals in $\mathbb{C}$.
- Creating a library of algebraic solutions by applying the results of this work to algebraic zeroes with absolute value smaller or equal than a certain value - referring to chapter 6.2 with $|k_0|$ analysis, extended by analysis of $k_4^2 - 2k_3$ in chapter 6.1.
- Replacing numeric algorithms in cases which can be determined exactly.
- Deriving solutions for polynomials of degree 6 or higher, with circular symmetries, leveraging the results of this research for quintics.
- Linking the results of this research - the resolvents of circular symmetric zeroes - to solvable Galois groups for these classes of polynomials and to Galois theory and other resolvent approaches. [1, 9]

8 References

[1] Wikipedia, *Quintic function*

[2] E.g. Univ. Augsburg, *Das Ikosaeder und die Gleichungen 5. Grades nach Felix Klein*

[3] Wikipedia, *Trigonometric constants expressed in real radicals*

[4] Wikipedia, *Algebraic number*

[5] Wikipedia, *Tschirnhaus transformation*

[6] Wikipedia, *Cubic equation*

[7] Wikipedia, *Quartische Gleichung*

[8] Wikipedia, *Lagrange Resolvente*

[9] Wikipedia, *Resolvent (Galois theory)*

[10] Wikipedia, *Cyclotomic polynomial*